Beginner's Guide to Streamlit with Python

Build Web-Based Data and Machine Learning Applications

Sujay Raghavendra

Apress®

Beginner's Guide to Streamlit with Python: Build Web-Based Data and Machine Learning Applications

Sujay Raghavendra
Dharwad, Karnataka, India

ISBN-13 (pbk): 978-1-4842-8982-2　　　　　ISBN-13 (electronic): 978-1-4842-8983-9
https://doi.org/10.1007/978-1-4842-8983-9

Managing Director, Apress Media LLC: Welmoed Spahr
Acquisitions Editor: Celestin Suresh John
Development Editor: Laura Berendson
Coordinating Editor: Mark Powers
Copy Editor: Kim Wimpsett

Cover designed by eStudioCalamar

Cover image by Sunbeam Photography on Unsplash (www.unsplash.com)

Distributed to the book trade worldwide by Apress Media, LLC, 1 New York Plaza, New York, NY 10004, U.S.A. Phone 1-800-SPRINGER, fax (201) 348-4505, e-mail orders-ny@springer-sbm.com, or visit www.springeronline.com. Apress Media, LLC is a California LLC and the sole member (owner) is Springer Science + Business Media Finance Inc (SSBM Finance Inc). SSBM Finance Inc is a **Delaware** corporation.

For information on translations, please e-mail booktranslations@springernature.com; for reprint, paperback, or audio rights, please e-mail bookpermissions@springernature.com.

Apress titles may be purchased in bulk for academic, corporate, or promotional use. eBook versions and licenses are also available for most titles. For more information, reference our Print and eBook Bulk Sales web page at www.apress.com/bulk-sales.

Any source code or other supplementary material referenced by the author in this book is available to readers on GitHub (https://github.com/Apress). For more detailed information, please visit www.apress.com/source-code.

Printed on acid-free paper

To
Prema Amma

Table of Contents

About the Author

Sujay Raghavendra is an IT professional with a master's degree in information technology. His research interests include machine learning, computer vision, NLP, and deep learning. He has been a consultant for multiple research centers at various universities. He has published many research articles in international journals and is the author of the book *Python Testing with Selenium*, published by Apress.

About the Technical Reviewer

Yanxian Lin is a data scientist with a PhD in biochemistry and molecular biology. He has seven years' experience with Python programming. He uses Streamlit for prototyping and demonstrating software developed in Python. Such software includes packages and machine learning models. He also uses Streamlit to develop graphical user interfaces for command-line tools. He is now an applied scientist in fraud detection.

Acknowledgments

I would like to thank my mom, Indumati Raghavendra, and brother, Sumedh Raghavendra, for their constant love, support, and encouragement.

Introduction

Streamlit is an application framework for web development based on Python. Streamlit was designed to reduce the time needed for developing web-based application prototypes for data and machine learning (ML) models. This framework helps to develop data enhanced analytics, build dynamic user experience, and showcase data for data science/ML models. The book provides insight that will help you to create high-quality applications using Python. The book uses a hands-on approach for creating such prototypes quickly.

The book starts by covering the basics of Streamlit by showing how to build a basic application and steps up incrementally, covering visualization techniques and their corresponding features. The book also covers various web application features that can be incorporated into Streamlit applications. You will also learn to handle the flow control of applications and status elements. The book also deals with performance optimization techniques necessary for data modules in an application. Last, you will learn how to deploy Streamlit applications on cloud-based platforms. You will also learn about a few prototype apps in the natural language processing (NLP) and computer vision fields that can be implemented from scratch using Streamlit.

By the end of this book, you will have a complete understanding of all the concepts, functionalities, and performance of Streamlit applications. This will enable you to be confident enough to develop your own dynamic Streamlit web applications.

What Is in the Book

The following is a brief summary of each chapter to help you better understand the structure of this book.

Chapter 1 will introduce you to the Streamlit library. It will walk you through the process of installing and running your first application.

Chapter 2 discusses text and dataframes that are represented in various forms in an application.

Chapter 3 covers visualization. Visualization is one of the important aspects of data science and machine learning. The visualizing techniques help you to understand the data more appropriately. In this chapter, we will implement the different visualizing techniques that are available in Streamlit as well as in other Python libraries for data science and machine learning developers.

Chapter 4 will discuss different media elements such as images, video, audio, etc., that can be implemented in an application.

Chapter 5 will introduce the button feature in Streamlit and how these buttons are used to select the required data to process or visualize data in applications being developed.

Chapter 6 mainly focuses on data provided by the user and how to process that data in an application. We will discuss user data in terms of forms.

Chapter 7 discusses column layouts, containers, and navigation in applications. It will also focus on how to switch between multiple pages in applications using navigation.

Chapter 8 introduces the custom handling of applications using control flow. You will also learn about status elements provided by Streamlit. Additionally, the chapter discusses how to handle huge data and optimize the performance of a Streamlit application.

Chapter 9 will introduce the development and deployment of an NLP application on the cloud-based platform Heroku.

Chapter 10 covers the development of a complete Streamlit application on a computer vision model from scratch. The features covered in the earlier chapters will be used in this application.

Who This Book Is For

This book is for professionals working in the data science and machine learning domains who want to showcase and deploy their work in a web application with no prior knowledge of web development. This book will help you learn basic Streamlit components in order to deploy your research work or prototypes. The book also guides you in developing dynamic data applications.

Source Code

The source code is available on GitHub at `https://github.com/apress/beginners-guide-streamlit-python`.

CHAPTER 1

Introduction to Streamlit

In this first chapter, we will take a look at Streamlit and its core features. We will also discuss why Streamlit is needed and compare it to alternatives. Next, we will discuss the installation steps to run Streamlit applications on various operating systems. Finally, we will develop and run our first Streamlit application.

What Is Streamlit?

Streamlit is an open-source Python framework released in October 2019. It acts as a medium between data and user interaction within an application. The goal of developing the Streamlit library was to enable data scientists and machine learning engineers to interpret and deploy data in a user interface (UI) with no prior knowledge of web development tools such as Flask, Django, Node, etc., using Python. Streamlit is the name of a software company founded by Adrien Treuille, Amanda Kelly, and Thiago Teixeira in 2018 in San Francisco, California.

The Streamlit framework uses the Python language and APIs for deploying built-in objects in the application. Thousands of companies, startups, professionals, and freelancers use Streamlit to deploy models or present data insights to their clients around the globe. We will now look into the reasons for using Streamlit.

© Sujay Raghavendra 2023
S. Raghavendra, *Beginner's Guide to Streamlit with Python*,
https://doi.org/10.1007/978-1-4842-8983-9_1

Why Streamlit?

The Streamlit library was developed to focus on prototype applications for data-driven and machine learning models. The prototype application is easy to use with no prior knowledge of web development. You just need basic knowledge of Python and Streamlit to develop such apps. Traditional development of applications is time-consuming when compared to Streamlit-based development, and Streamlit apps are interactive in nature. An application developed using Streamlit will act as a tailwind for many companies by reducing process time and cost.

Why Streamlit for Data Science and ML Engineers?

We know that data science and machine learning use a technical medium to extract data and present this information with its associated insights to clients/stakeholders. Most of the time Python scripts or programs are developed as custom tools by data scientists and ML engineers. When they want to showcase these insights with their clients/stakeholders, it becomes extremely difficult to share the developed scripts/programs. It is also difficult to represent the data to nontechnical customers/users.

Let's assume that you are a data science or ML employee working in a company. With the client's requirements, you have been asked to develop a customized interactive tool so users can explore data along with associated visualization. When the client wants to check the application being developed, you can use the Streamlit library to develop a customized interactive application and host it on a server to share with your client/stakeholder.

Now you know the perspective of data scientists and ML engineers for developing a Streamlit application. We will now look at the major features of Streamlit in the next section.

Features of Streamlit

We will discuss a few important features that make Streamlit unique compared to other tools and applications.

Open Source

Streamlit is open source, and hence it is free to use. A large community of data scientists and ML engineers use it for developing their applications.

Platforms

Streamlit can run on any platform, which means once applications are developed, they can be modified or altered on any other system.

Ease of Development

You can develop dashboards and data- or ML-driven applications with user interaction in no time as less code is needed.

Interactive Applications

The user interacts with the application by modifying or uploading data to visualize it or apply the implemented model data. You will see the various user interactions that are possible in applications using Streamlit in upcoming chapters.

Reduced Time of Development

The development time for Streamlit applications is less than with various competitors in the domain. The autoreloading feature previews the changes made to the application, reducing the time needed.

No Core Web Development Knowledge

The Streamlit library binds the front end and back end of an application. The developer does not need any prior web development knowledge to develop data-driven applications as Streamlit takes care of it.

Easy to Learn

You need some basic knowledge of the Python programming language along with Streamlit's syntaxes that will be discussed in this book to build data-driven applications.

Model Implementation

You can implement pretrained models in an application and can analyze the model results. This type of application may be client-specific or testing-specific. The Face-GAN explorer is one such example of a model implementation.

Compatibility

Streamlit supports various Python-driven libraries for visualization, computer vision, machine learning, and data science. Table 1-1 lists some of them.

Table 1-1. *Python Libraries Supported by Streamlit*

Frameworks	Python Libraries
Visualization	Matplotlib, Plotly, Seaborn, Bokeh, Altair
Machine learning	PyTorch, TensorFlow, Keras, Sci-kit Learn
Data science	Numpy, Pandas
Computer vision	OpenCV, imgaug, SimpleCV, BoofCV

Literate Programming Document

In a Streamlit application, you can parse any language such as Markdown, LaTeX, etc., in a .py file and render it as HTML, which makes a .py file act as a literate programming document where the back end is hosted by a Python-based server. You can run any programming language by defining the name in a .py file; this will be discussed in later chapters of this book.

Streamlit Cloud

The developed applications can be deployed and shared using the Streamlit Cloud (https://streamlit.io/cloud) by giving access to clients or team members remotely. (There are no mandatary restrictions to use the Streamlit Cloud for the applications developed.)

Optimize Change

The changes made in your Python script files are displayed directly in the application, making development easier. You will learn more about this when you create your first application later in this chapter.

Error Notifications

If there are any errors that have incurred in our code, they are displayed in the application instead of at the command prompt, making error detection easier. The errors may be from Streamlit's built-in functions or any library that has been used for developing the application.

Comparing Streamlit to Alternative Frameworks

Now that you know about the features of Streamlit, we will now compare the Streamlit features to its competitors; see Table 1-2.

Table 1-2. *Comparing Streamlit to Its Competitors*

	Streamlit	Panel	Dash	Voila
Owner	Streamlit	Anaconda	Plotly	Quantstack
Release year	2019	2018	2017	2019
License type	Apache	BSD	MIT	BSD
Front end	React	HTML template, Bokeh, and any JavaScript library	React	HTML template or Jupyter
Back end	Tornado	Tornado	Tornado	Flask
Public deployment	Any cloud environment (also provides the Streamlit Cloud for easy connection and deployment)	Any cloud environment	Any cloud environment	Any cloud environment
Complex apps	Supports multipage	Supports by configuring reactive components	Supports but requires manual event handling	Difficult to handle complex apps
Hosting service support	Streamlit Cloud	Jupyterhub/ Binder	Need to purchase Enterprise Edition	Jupyterhub/ Binder
Product support	Open-source community	Open-source community	Open-source community/ business support	Open-source community

You now have a complete overview of Streamlit and understand why it was developed and its associated features. Next we will get one step closer to creating our first Streamlit application by installing Python on our system.

Installing Python

The first step to get started is to install Python. Follow these steps to install Python in Windows:

1. Open any browser and navigate to `https://python.org/downloads/windows/link`.

2. Under Python Release for Windows, choose a Python version above 3.7 to download as Streamlit supports Python versions 3.7, 3.8, 3.9, and 3.10 (the latest version is 3.10.4).

3. Select your Windows version (there is a Windows executable installer for 32-bit or 64-bit).

4. Finally, run the downloaded installer and follow the installer steps; also remember update the ADD PATH variable.

Note   Python is built into Mac and Linux OS. Check the version using the command `python3 --version` and upgrade if the Python version is lower than 3.7.

Now we will see how to install the Streamlit library for various operating systems.

Installing Streamlit on Windows

Once we have installed Python in our system, we can now utilize the `pip` command to download the Streamlit package. Use the following the `pip` command to install Streamlit:

```
pip install streamlit
```

Installing Streamlit on Linux

Assuming we have already installed a Python version higher than 3.7 on our system, we need install `pip` to directly install our Streamlit library. To install it, we will use the following command:

```
sudo apt - get install python3 - pip
```

We will now install the Streamlit library.

```
pip install streamlit
```

Installing Streamlit on macOS

To install Streamlit on a Mac system, first we need to install `pip` similar to what we have done for Linux. The command to install `pip` is given here:

```
sudo easy_install pip
```

We will now download the Streamlit library on macOS by using the following command:

```
pip install streamlit
```

Note   Currently, the Streamlit library supports Python versions ranging from 3.7 to 3.10 only. We have used Python version 3.8.12 in this book.

Testing the Streamlit Installation

We will now check whether the Streamlit library has been installed by using the Streamlit command. Open the command prompt and type the following command, as shown in Figure 1-1:

```
streamlit hello
```

Figure 1-1. *Command prompt screen displaying Streamlit application running on local server*

After running the command, we will use Streamlit to run a built-in Welcome application in our default web browser at the localhost address, as displayed in Figure 1-2. In our case, the localhost address is `http://localhost:8052`.

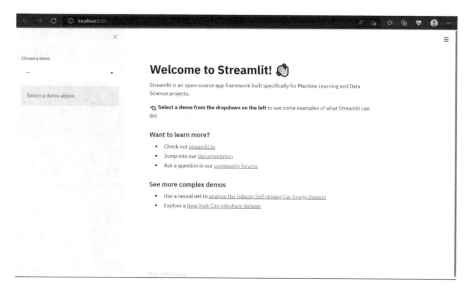

Figure 1-2. *Displaying default Streamlit page*

There are a few demo applications that show animation, plotting, mapping, and dataframes available in the drop-down menu of the Welcome screen. When we select Mapping Demo from the drop-down menu, then the app and code are generated, as shown in Figure 1-3.

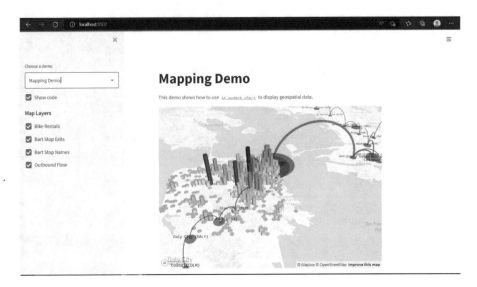

Figure 1-3. *Displaying a demo application*

Now we will discuss how to create our first application in Streamlit.

Creating Our First App

Assuming that you have already installed the Streamlit Python library as per the instructions, you will see step-by-step instructions for how to create your first Streamlit application.

At first, you need to create a new Python file (.py) in any of the specified directories.

Next, we will write code to import the Streamlit library. Choose any code editor you'd like. I have used Visual Studio for writing the following script:

```
import streamlit as st
```

We will display "Hello" text in our application using the write function; this is similar to the print function in the Python programming language.

```
st.write("Hello")
```

Finally, we will save our file and run the following command in our command prompt from the directory where we have saved it:

```
streamlit run filename.py
```

Now, we will see only "Hello" text in our application, as shown in Figure 1-4.

Figure 1-4. *Displaying "Hello" text in the created application*

Here, the filename is the name given to our Python script, and it will display our "Hello" text when we open our default browser. We will now write one more `write` statement with "World" as the text.

```
st.write('World!!!!')
```

When we make any changes to our created Python file, the application running in the browser will show the commands in Figure 1-5 in the top-right corner of the application.

Figure 1-5. *Displaying buttons in an application when the Python script is edited/altered*

"Source file changed" is the message shown when we save any changes made in our Python script and do not refresh the page. The Rerun button will run all the code blocks from the start and display the changes thereafter.

When we choose the "Always rerun" button in the application, the application automatically optimizes with the changes made and displays the changes.

Listing 1-1 shows the complete program for creating our first application.

Listing 1-1. 1 _hello_world.py

```
#Import Streamlit library
import streamlit as st

#Displaying Plain Text
st.write("Hello")
st.write('World!!!!')
```

Figure 1-6 shows the screen after refreshing or rerunning the file for the previous script.

Figure 1-6. *Displaying "Hello World!!!" in an application*

We have developed a basic Streamlit application displaying the "Hello World!!!" text in it. This is to give you an idea of creating Python scripts using Streamlit.

Summary

In this chapter, we got a complete overview of the Streamlit application by discussing its usage for data scientists and ML engineers. We discussed its core features and compared it to other tools available. Then we talked about installing Python versions that Streamlit is compatible with and went through the instructions to install Python and Streamlit on different platforms. Finally, we created our first basic Streamlit application using the `streamlit` command in a Python script.

In the next chapter, we will discuss various ways in which textual data is represented in an application. We will also cover more about the data elements that are necessary to develop data-centric applications.

CHAPTER 2

Text and Table Elements

In the first chapter, you gained a complete overview of Streamlit's core features and how it is different from its competitors in the data science and machine learning domain. We also discussed the installation steps that are necessary on various platforms and created our very own Streamlit application.

In this chapter, we will discuss not only on how we can format text in different ways, such as using Markdown, captions, blocks of code, and LaTeX, but also how to add title, headers, and subheaders in the app we have created. Next, we will look at table data elements and how we can display them in our application.

Let's start by looking at the various text elements available in the Streamlit library.

Text Elements

We define our text elements as objects with descriptions that are used in Streamlit applications as title, headers, subheaders, captions, and plain text. We will begin by looking at titles as our first text element.

© Sujay Raghavendra 2023
S. Raghavendra, *Beginner's Guide to Streamlit with Python*,
https://doi.org/10.1007/978-1-4842-8983-9_2

Titles

We will begin by writing a title in our Streamlit application. The title is case-sensitive, and it will be represented on each document, but this is not enforced. In other words, the title is similar to a heading of a section. The code in Listing 2-1 will yield the output in Figure 2-1.

Listing 2-1. 01_title.py

```
# Import Streamlit library
import streamlit as st

# Writing Title
st.title("This is our Title")
```

Figure 2-1. *Displaying a title*

We can see the text, shown in Figure 2-1, that we have mentioned in the st.title() function. We can also add an anchor in our title to navigate to another link or page. (We will learn more about this in Chapter 7.) As we have not added an anchor to our title example, the anchor will be generated using the same text that we have written for the title, i.e., *This is our Title*. We will now see how to add an anchor to our title; see Listing 2-2.

Listing 2-2. 02_title_anchor.py

```
# Import Streamlit library
import streamlit as st

# Writing Title with ANCHOR
st.title("This is our Title", anchor="Apress")
```

In Listing 2-2, we have used Apress as our anchor. This is used to link pages or a few sections in a page to another page or sections within the same page. The anchor functionality is optional and can also be set to None if we are not using it.

In the next sections, we will study how to define headers and subheaders using the Streamlit library, respectively.

Headers

We all know that a header is used to represent a set of content in an application. We will now define the built-in header function available in the Streamlit library. See Listing 2-3.

Listing 2-3. 03_header.py

```
# Import Streamlit library
import streamlit as st

# Header
st.header("""This is our Header""")
```

Figure 2-2 shows our header text displayed in our application. Now, let's check out subheaders in the next section.

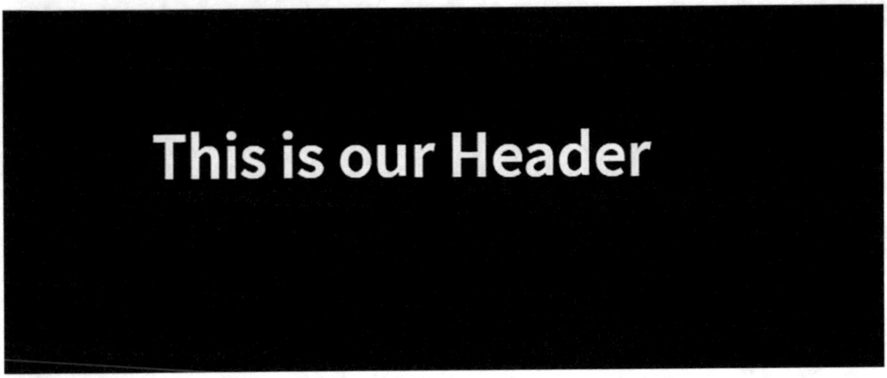

Figure 2-2. *Displaying a header*

Subheaders

As the name suggests, a subheader is a subtype of header. It is written after the header in a smaller font size than the header. Listing 2-4 shows our example, and the associated output is in Figure 2-3.

Listing 2-4. 04_sub_header.py

```
# Import Streamlit library
import streamlit as st

# sub-header
st.subheader("""This is our Sub-header""")
```

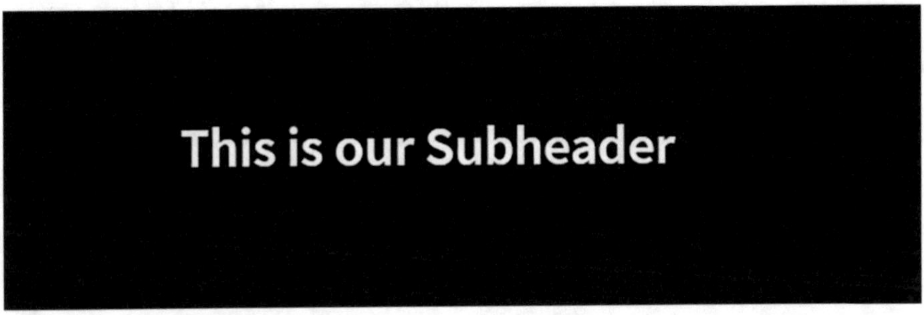

Figure 2-3. *Displaying subheader*

We can also apply an anchor parameter as we have done for the st.
title() function.

Note   We can use an anchor parameter in st.title(), st.
header(), and st.subheader(), but it is optional.

Captions

A caption is usually associated with a description in the application. The
size of a caption is small; see Listing 2-5 and Figure 2-4.

Listing 2-5. 05_caption.py

```python
# Import Streamlit library
import streamlit as st

# Caption
st.caption("""This is our Caption""")
```

Figure 2-4. *Displaying a caption*

We know that caption text is an explanation that describes notes,
footnotes, tables, images, and videos.

Let us now compare all three types of text (titles, headers, and
captions) in one go to understand the difference and see how they are

depicted in the Streamlit application. See Listing 2-6 and the output in Figure 2-5.

Listing 2-6. 06_comparison.py

```
# Import Streamlit library
import streamlit as st

# Title
st.title("""This is our Title""")

# Header
st.header("""This is our Header""")

# Sub-header
st.subheader("""This is our Subheader""")

# Caption
st.caption("""This is our Caption""")
```

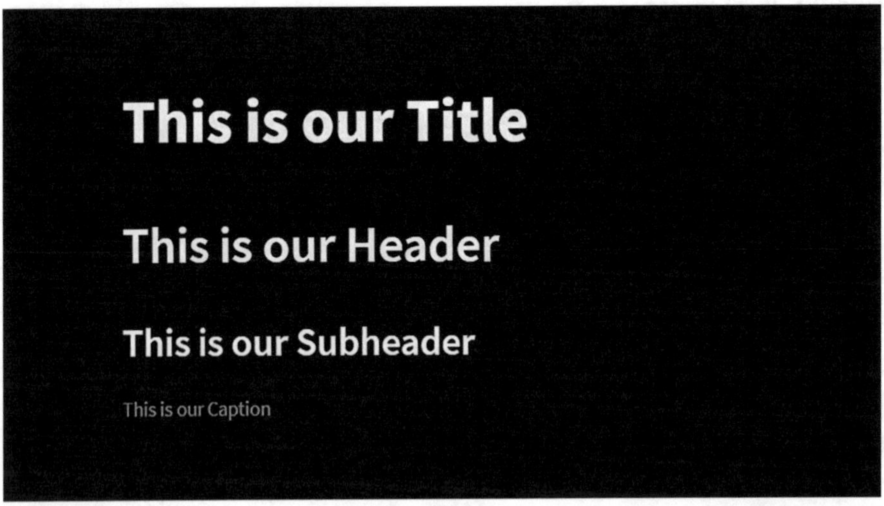

Figure 2-5. *Comparing the title, header, subheader, and caption elements by displaying them in one page*

Now that we know the difference between the three, we will move on to write text in plain format.

Plain Text

Let's display some plain text in our application. The syntax for it is given here:

st.text("your text")

This st.text() function will display any text that is written in between the quotes. The syntax follows the same string procedures used in Python. The text may be written in single or double or triple quotes, as shown in Listing 2-7. Figure 2-6 shows the output.

Listing 2-7. 07_plain_text.py

```
#Import Streamlit library
import streamlit as st

#Displaying Plain Text
st.text("Hi,\nPeople\t!!!!!!!!!!")
st.text('Welcome to')
st.text(""" Streamlit's World""")
```

Figure 2-6. *Displaying plain text*

This is how plain text is displayed in the application we have created. The screen is in dark mode, so we are seeing our application with a black background.

Note   Each time we write the `st.text()` function, it will appear on a new line in an application.

In the next section, we will discuss how to create Markdown text from plain text.

Markdown

Markdown is a markup language used to create formatted text. We will now see how to display plain text in Markup language in our application. See Listing 2-8 and the output in Figure 2-7.

```
st.markdown()
```

Listing 2-8. 08_markdown.py

```
#Import Streamlit library
import streamlit as st

#Displaying Markdown
st.markdown("# Hi,\n# ***People*** \t!!!!!!!!!")
st.markdown('## Welcome to')
st.markdown("""### Streamlit's World""")
```

Figure 2-7. *Displaying Markdown-formatted text*

The text written in the `st.markdown()` function should be in the Markup language in order to get the formatted text shown in Figure 2-7.

LaTeX

LaTeX is formatted text used mainly for technical documentation. Streamlit supports text written in LaTeX format. See Listing 2-9 and the output in Figure 2-8.

Listing 2-9. 09_latex.py

```
#Import Streamlit library
import streamlit as st

#Displaying Latex
st.latex(r'''cos2\theta = 1 - 2sin^2\theta''')
st.latex("""(a+b)^2 = a^2 + b^2 + 2ab""")
st.latex(r'''\frac{\partial u}{\partial t}
   = h^2 \left( \frac{\partial^2 u}{\partial x^2}
      + \frac{\partial^2 u}{\partial y^2}
```

```
+ \frac{\partial^2 u}{\partial z^2} \right)''')
```

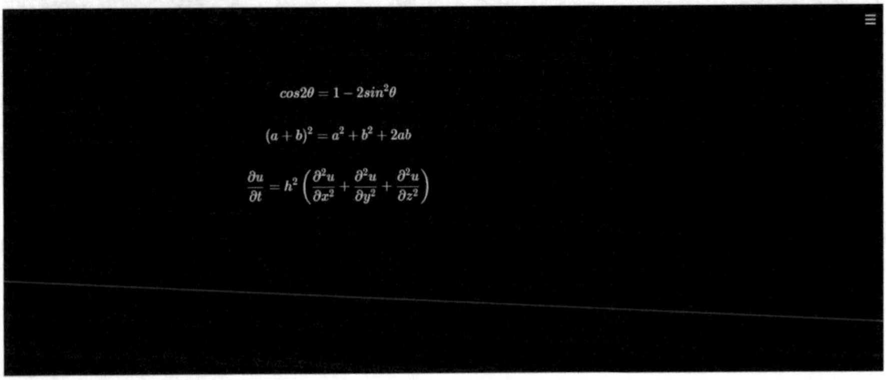

$$cos2\theta = 1 - 2sin^2\theta$$

$$(a + b)^2 = a^2 + b^2 + 2ab$$

$$\frac{\partial u}{\partial t} = h^2 \left(\frac{\partial^2 u}{\partial x^2} + \frac{\partial^2 u}{\partial y^2} + \frac{\partial^2 u}{\partial z^2} \right)$$

Figure 2-8. *Displaying mathematical equations using the LaTeX format*

Code

The Streamlit library offers some flexibility to display the code of various programming languages. The text is highlighted as shown in a code editor. See Listing 2-10 and the output in Figure 2-9.

Listing 2-10. 10_code.py

```python
# Import Streamlit library
import streamlit as st

# Displaying Python Code
st.subheader("""Python Code""")
code = '''def hello():
    print("Hello, Streamlit!")'''
st.code(code, language='python')
```

```
# Displaying Java Code
st.subheader("""Java Code""")
st.code("""public class GFG {
    public static void main(String args[])
    {
        System.out.println("Hello World");
    }
}""", language='javascript')

st.subheader("""JavaScript Code""")
st.code(""" <p id="demo"></p>

<script>
try {
  adddlert("Welcome guest!");
}
catch(err) {
  document.getElementById("demo").innerHTML = err.message;
}
</script> """)
```

If we do not add the programming language for the code specified, then the code will appear in an unstyled format, like our code in the HTML language. The language option is not mandatory unless there is more than one language used. We can also copy the code written in the application with a shortcut appearing at top-right corner of the code box.

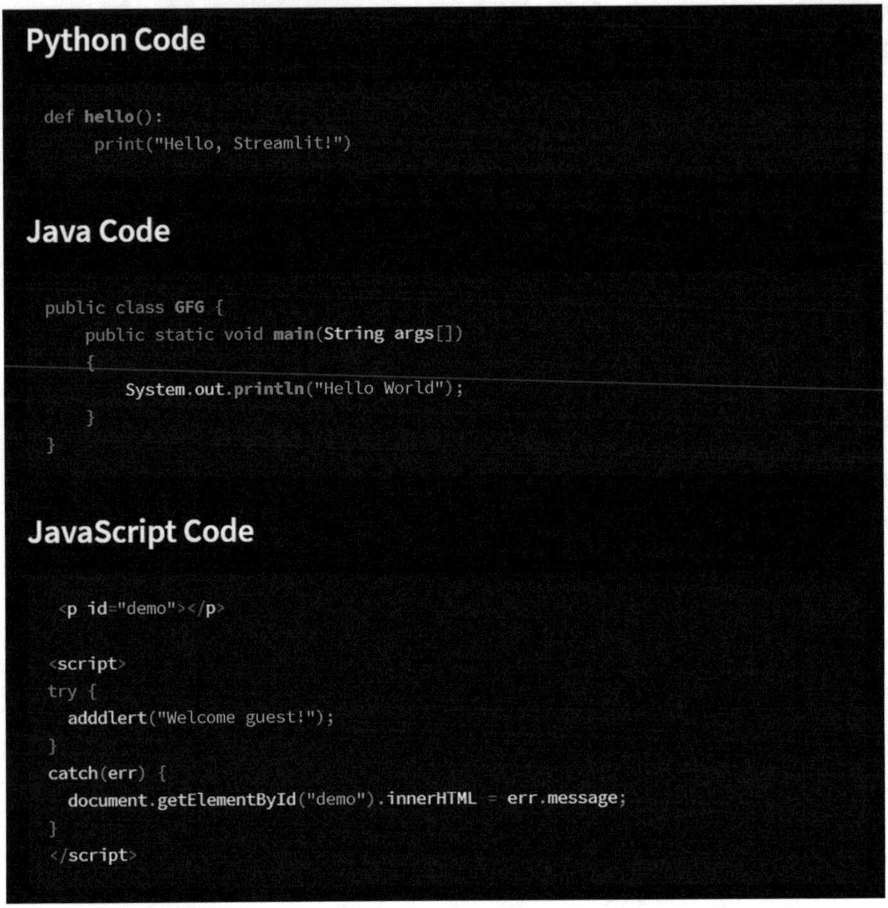

Python Code

```python
def hello():
    print("Hello, Streamlit!")
```

Java Code

```java
public class GFG {
    public static void main(String args[])
    {
        System.out.println("Hello World");
    }
}
```

JavaScript Code

```javascript
<p id="demo"></p>

<script>
try {
  adddlert("Welcome guest!");
}
catch(err) {
  document.getElementById("demo").innerHTML = err.message;
}
</script>
```

Figure 2-9. *Displaying a block of code*

Now, we will move on to other elements known as *data elements*. This is one of the best features in an Streamlit application to help define data for exploration.

Data Elements

We have seen how to represent textual data in various formats in our Streamlit application. We will now move on to the various data forms that are available to display in the application. Let's look at the data elements one by one.

Dataframes

As we know, a dataframe is one of the most important data formats used in data science and machine learning projects. It consists of two-dimensional data with rows and columns that act as labeled data. The Streamlit library displays a dataframe in an interactive table format, which means the table is scrollable and can also change table size dynamically. See Listing 2-11 and the output in Figure 2-10.

Listing 2-11. 11_dataframe.py

```
# Import Necessary libraries
import streamlit as st
import pandas as pd
import numpy as np

# defining random values in a dataframe using pandas and numpy
df = pd.DataFrame(
    np.random.randn(30, 10),
    columns=('col_no %d' % i for i in range(10)))

st.dataframe(df)
```

	col_no 0	col_no 1	col_no 2	col_no 3	col_no 4	col_no 5	col_no 6	col
0	-0.6434	1.1279	-1.9456	0.6087	-0.0176	1.6158	0.8694	0
1	0.0936	0.9285	2.0637	0.1619	-0.2379	-2.7233	-0.1078	-2
2	0.5882	0.6864	-0.3826	0.3382	-0.2125	-1.0974	0.3424	-0
3	1.3862	-0.3612	-0.4308	-0.0512	0.2785	-0.5771	-0.1945	0
4	-1.1207	1.9191	-1.3120	-0.3377	1.0766	-0.4635	0.1325	-0
5	1.5722	0.2076	-1.5298	-2.3750	-1.6807	0.5050	0.1316	-0
6	-0.4158	0.2905	-0.5437	0.4439	-0.0174	0.5900	-0.4961	0
7	0.6335	-1.2750	-0.7651	0.0717	-0.6184	1.1031	-0.7191	-0
8	1.0381	-1.2384	0.5992	-1.1823	-1.3271	1.1418	-0.3872	1
9	-0.6563	-0.4354	-0.4440	0.3892	-0.1721	0.6626	0.9434	

Figure 2-10. *Displaying a dataframe with scroll and filter options*

We can fix the height and width of the dataframe that is displayed in an application. The default size that is determined by Streamlit is static and will not be affected even when we change the size of the table from within the application.

Note   This is also known as a *dynamic table* as we can change the width and height of the column in a table from the application directly with no code changes.

We can highlight dataframe objects using minimum, maximum, and null values by using the highlight function. See Listing 2-12.

Listing 2-12.   12_dataframe_highlight.py

```
import streamlit as st
import pandas as pd
import numpy as np
```

```
# defining random values in a dataframe using pandas and numpy
df = pd.DataFrame(
    np.random.randn(30, 10),
    columns=('col_no %d' % i for i in range(10)))
# Highlighting minimum value objects
st.dataframe(df.style.highlight_min(axis=0))
```

In Listing 2-12, we have highlighted the objects with minimum values in the dataframe. Figure 2-11 shows the application displaying highlights for the minimum values in the dataframe.

	col_no 0	col_no 1	col_no 2	col_no 3	col_no 4	col_no 5	col_no 6	↘
0	1.178489	0.370286	0.585741	-2.282765	1.237429	0.657911	-0.305897	
1	-0.007425	0.126586	-0.395297	-0.202437	-0.836234	1.504649	0.005744	
2	-0.919069	1.412931	-0.886233	1.565651	0.181077	0.395909	0.315348	
3	0.232683	0.335672	0.309013	-1.989317	-0.485276	-0.466041	-0.534550	
4	0.102847	-0.275711	0.994698	0.546669	-0.869790	0.211362	0.878219	
5	-0.693490	-0.028150	0.353265	1.896758	1.237518	-0.515882	0.325412	
6	-0.857123	1.364154	0.321079	-0.639655	-0.072768	0.067078	-0.340984	
7	0.715961	0.826992	0.799698	1.674534	-0.208510	1.703436	0.719906	
8	-0.020842	-0.864880	-0.889566	0.959570	-1.271859	-1.593557	-0.483050	
9	-1.917296	-0.373501	-1.000166	1.573044	0.212838	-1.380918	0.694537	

Figure 2-11. *Displaying highlighted objects in a dataframe*

Note   Copying data to the clipboard is not yet supported in the Streamlit Cloud.

We will learn how to display a table in the next section.

Tables

As we know, tables and dataframes are similar in nature, both having rows and columns. But a table displayed in Streamlit is static in nature and is displayed directly on the application page. See Listing 2-13 and the output in Figure 2-12.

Listing 2-13. 13_table.py

```
import streamlit as st
import pandas as pd
import numpy as np

# defining random values in a dataframe using pandas and numpy
df = pd.DataFrame(
    np.random.randn(30, 10),
    columns=('col_no %d' % i for i in range(10)))
# defining data in table
st.table(df)
```

	col_no 0	col_no 1	col_no 2	col_no 3	col_no 4	col_no 5	col_no 6	col_no 7	col_no 8	col_no 9
0	-0.1711	-1.2524	-0.4512	-1.0797	-1.2391	-0.2739	1.3886	0.8321	-0.8045	2.0818
1	-1.1906	-0.6553	2.0707	0.3506	-0.9852	1.3927	0.6419	-1.1691	-0.7666	-0.7156
2	1.1270	-0.0169	0.0618	0.7021	0.7420	-0.5694	1.5728	-0.1863	-0.0405	-0.4008
3	0.3998	-0.1301	-0.8599	-1.1035	0.8917	-0.5313	-0.4153	0.2817	-0.6096	0.2667
4	0.3498	-0.4242	-1.1024	-0.0181	-0.6464	-1.2974	-1.5213	0.6771	0.6180	0.8809
5	1.2833	-0.4015	1.1139	-0.3152	-1.1422	-0.0401	0.0187	1.8377	-0.3921	0.1077
6	0.1062	-0.2024	-0.8603	-1.0518	0.2840	2.0415	1.2829	1.1787	-0.4589	0.3597
7	0.3885	-0.1468	0.5383	0.9655	1.8012	0.4044	-0.2817	-0.9778	-0.3329	-0.6630
8	0.4865	0.8618	-0.8024	0.2819	3.8101	0.1344	-0.1674	0.3116	-0.0301	-0.4996
9	0.6758	-1.9253	1.2721	-0.7162	-0.3922	-0.7738	0.5455	-0.2882	0.1300	0.1479
10	0.7633	-0.3913	-0.4861	-0.2736	-2.1197	-0.0117	0.2818	-1.6780	0.3893	-0.6213
11	-1.6006	0.1852	-1.1987	2.1227	0.6137	1.7713	-1.9606	-0.6920	1.6261	-0.0929
12	0.7706	-1.2332	1.0316	-1.1568	0.2338	-0.2399	2.0613	0.9389	-0.1127	-0.0632
13	0.4648	1.4910	0.5468	0.9921	0.7015	-0.6781	1.3855	-2.0813	-1.5417	-0.7542
14	-0.0130	0.2863	-0.9775	2.6328	-1.0912	1.3829	-0.3959	0.7911	1.5991	0.1396
15	1.5441	0.2881	1.2490	-0.2601	0.8950	0.3614	-0.4261	0.2931	0.1244	0.2683
16	1.0462	1.0031	-0.1647	-0.5131	0.6060	-0.7418	-1.0485	0.8143	0.0507	-0.7106
17	0.4476	0.4669	0.2582	-0.5129	1.0498	1.5474	0.4907	0.1669	2.6911	0.5556
18	0.5693	-0.0007	1.0249	-0.2031	1.4196	-0.0425	0.6408	0.3051	-0.0107	-1.9977
19	0.7333	1.1871	-0.2518	0.9659	-0.2808	0.7401	-1.2950	0.0363	0.2023	0.5388

Figure 2-12. *Displaying data in a table format*

As we can see in Figure 2-12, the complete table is displayed in the application instead of the default data that is displayed using a dataframe. This is one of the major differences between a table and a dataframe.

Note   The table defined in the Streamlit application is static in nature, and there is no option for scrolling and changing the dimensions of the table within the UI like we saw with dataframes.

Next, we will look at one more data element in which we can display data in a unique way.

Metrics

In a Streamlit application, we can display data with indicators using the metric() function. This indicator helps the user see any changes in the data easily. We can define the change in data whether it is positive or negative or neutral.

We will now display temperature data in our application by using the metric() function. See Listing 2-14 and the output in Figure 2-13.

Listing 2-14. 14_metric.py

```
import streamlit as st

# Defining Metrics
st.metric(label="Temperature", value="31 °C", delta="1.2 °C")
```

The label is Temperature, the value is 31, and the delta is 1.2 degrees Celsius. The delta shows the change in variation application whether it is a positive or negative or null value associated with the label. See Figure 2-13.

Figure 2-13. *Displaying temperature metrics*

Similarly, we can display any metrics such as speed, rainfall, energy, calories, etc., in the label by assigning application the value and delta value. See Listing 2-15.

Listing 2-15. 15_metric_all.py

```python
import streamlit as st

#Defining Columns
c1, c2, c3 = st.columns(3)

# Defining Metrics
c1.metric("Rainfall", "100 cm", "10 cm")

c2.metric(label="Population", value="123 Billions", delta="1
Billions", delta_color="inverse")

c3.metric(label="Customers", value=100, delta=10, delta_
color="off")

st.metric(label="Speed", value=None, delta=0)

st.metric("Trees", "91456", "-1132649")
```

In Listing 2-15, we can see how different metrics formats application are supported by Streamlit. The delta is the value that helps users to know if any changes have been incurred from the value defined. The delta_ color is the color defined for positive and negative values; here we use green and blue, respectively. See Figure 2-14.

Figure 2-14. *Displaying metrics in various forms*

We can also change the delta indicator application to be inversed or neutral by turning it off. This is done in the following JSON example.

JSON

We know that tabular data application in data science is sometimes in JSON format. JSON format is a way to interchange data. It is easy to read and write for users. When an object is a string, we will assume that it will contain serialized JSON. Listing 2-16 shows how a string or object is displayed as a JSON string in Streamlit.

Listing 2-16. 16_json.py

```
import streamlit as st

#Defining Nested JSON
st.json
(
  { "Books" :
```

```
[{
"BookName" : "Python Testing with Selenium",
"BookID" : "1",
"Publisher" : "Apress",
"Year" : "2021",
"Edition" : "First",
"Language" : "Python",
},
{
    "BookName": "Beginners Guide to Streamlit with Python",
    "BookID" : "2",
    "Publisher" : "Apress",
    "Year" : "2022",
    "Edition" : "First",
    "Language" : "Python"
}]
}
)
```

Here we have used nested JSON in the code. Figure 2-15 shows how it is displayed in the application.

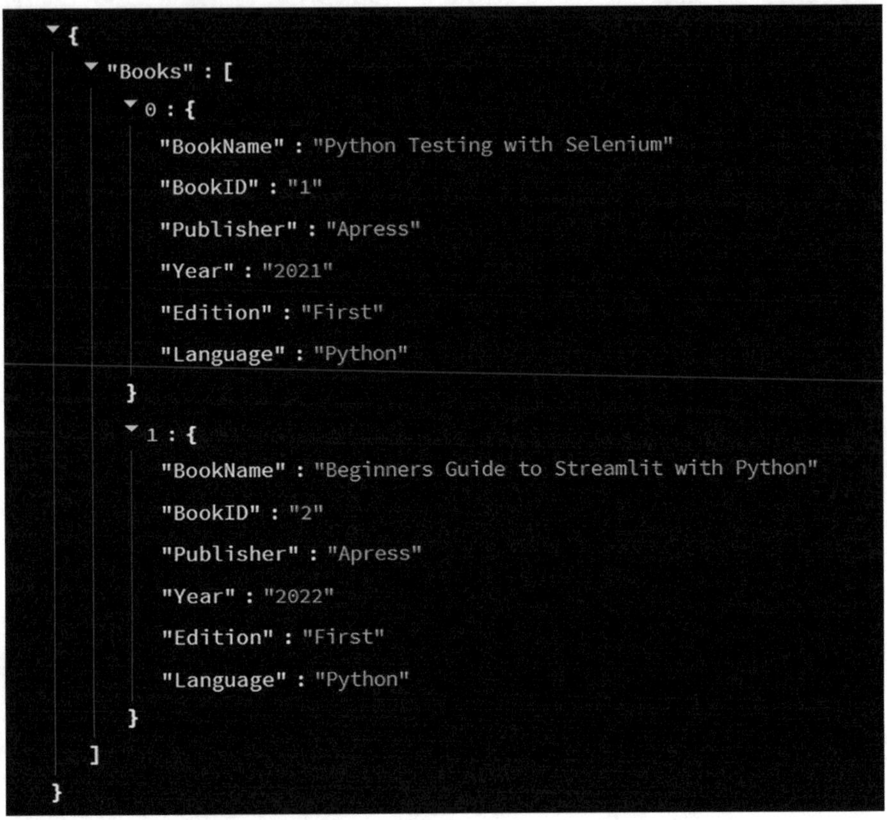

```
▼ {
    ▼ "Books" : [
        ▼ 0 : {
            "BookName" : "Python Testing with Selenium"
            "BookID" : "1"
            "Publisher" : "Apress"
            "Year" : "2021"
            "Edition" : "First"
            "Language" : "Python"
        }
        ▼ 1 : {
            "BookName" : "Beginners Guide to Streamlit with Python"
            "BookID" : "2"
            "Publisher" : "Apress"
            "Year" : "2022"
            "Edition" : "First"
            "Language" : "Python"
        }
    ]
}
```

Figure 2-15. *Displaying JSON data*

Up to now, we have seen how to display all types of data formats in the Streamlit application. Now we will check one more method that will also display data in an application.

The write( ) Function as a Superfunction

The write() function is similar to the print() function in Python or the printf() statements in C and Java. The write() function is an output function that is displayed on the screen. The basic syntax is given here:

```
st.write(input)
```

There is twist in the write() function in that it does more than just display text. We will see how the write() function is used in different ways. First, we will implement a dataframe using the write() function. See Listing 2-17.

Listing 2-17. 17_write_df.py

```
import streamlit as st
import pandas as pd

# Dataframe in write function
st.write(pd.DataFrame({
    'column one': [5.436, 6.372, 3.645, 4.554, 7.263],
    'column two': [99, 55, 75, 41, 37],
 }))
```

In Figure 2-16, we have embedded a dataframe in a write() function. Similarly, we can induce multiple arguments in a write()function.

	column one	column two
0	5.4360	99
1	6.3720	55
2	3.6450	75
3	4.5540	41
4	7.2630	37

Figure 2-16. *Displaying a dataframe using a write() function*

We will now see one more example with multiple arguments embedded in a write() function. See Listing 2-18.

Listing 2-18. 18_write_mul_args.py

```python
import streamlit as st
import pandas as pd
import numpy as np

df = pd.DataFrame(
    np.random.randn(30, 10),
    columns=('col_no %d' % i for i in range(10)))

# defining multiple arguments in write function
st.write('Here is our Data', df, 'Data is in dataframe
format.\n', "\nWrite is Super function")
```

We have not used a default dataframe definition in the write()
function. Here the write() function automatically checks for the value
corresponding to the data type and displays it in our application, as shown
in Figure 2-17.

Here is our Data

	col_no 0	col_no 1	col_no 2	col_no 3	col_no 4	col_no 5	col_no 6	col
0	-0.0826	0.9812	-1.0660	-0.6651	-0.5748	1.0362	-0.7820	-0
1	2.1039	1.3708	-0.1731	0.0603	-1.0001	0.0941	0.1091	-1
2	0.5453	1.5550	-1.4294	0.3066	-0.8883	-0.3546	-0.7812	0
3	-1.4679	1.0498	0.9491	-1.7884	-1.6109	-0.7189	-0.8311	-0
4	0.3405	-0.2329	-0.0060	-0.5866	1.2450	0.2760	0.8442	-1
5	-1.7706	-1.8556	0.5666	-1.1773	-0.0011	0.0118	0.5512	0
6	0.1757	0.7714	1.5473	0.5968	-0.0763	-0.2838	-0.9376	1
7	-0.4545	1.2642	0.0309	0.1742	-1.4680	-0.3182	-0.5456	0
8	1.2006	0.9102	1.4066	0.6649	0.9911	0.5392	0.4832	2
9	0.9138	-0.3391	1.9785	-0.1932	-0.4770	-1.1909	-1.1335	0

Data is in dataframe format.

Write is Super function

Figure 2-17. *Displaying multiple objects defined in a write() function*

In this example, we have used two data types; one is text, and the other is a dataframe in a single `write()` function, which is shown in Figure 2-17.

Finally, we will see how we can use the `write()` function to display a simple chart in our application. See Listing 2-19 and the output in Figure 2-18.

Listing 2-19. 19_write_chart.py

```
# importing Necessary Libraries
import pandas as pd
import numpy as np
import altair as alt
import streamlit as st

# Defining random Values for Chart
df = pd.DataFrame(
    np.random.randn(10, 2),
    columns=['a', 'b'])

# Defining Chart
chart = alt.Chart(df).mark_bar().encode(
    x='a', y='b',  tooltip=['a', 'b'])

# Defining Chart in write() function
st.write(chart)
```

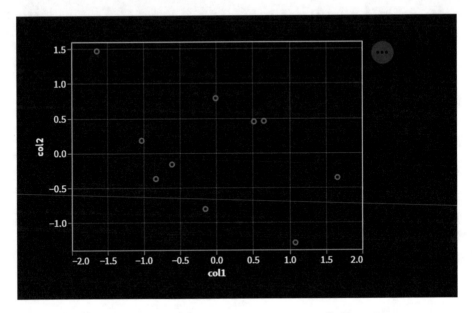

Figure 2-18. *Displaying a chart using the write() function*

Finally, we will check out the magic feature in Streamlit.

Magic

Magic is a feature that enables us to define data elements in our application without explicitly stating any command. We will see how the magic works in Listing 2-20 and the output in Figure 2-19.

Listing 2-20. 20_magic_feature.py

```
# Math calculations with no functions defined
"Adding 5 & 4 =", 5+4

# Displaying Variable 'a' and its value
a = 5
'a', a
```

```
# Markdown with Magic
"""

# Magic Feature

Markdown working without defining its function explicitly.
"""

# Dataframe using magic
import pandas as pd
df = pd.DataFrame({'col': [1,2]})
'dataframe', df
```

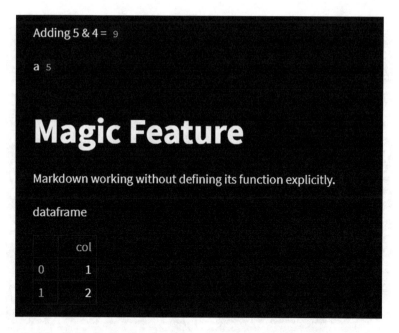

Figure 2-19. *Displaying a dataframe using a magic function*

We can even display a chart in our application using the magic feature, as shown in Figure 2-20 using Listing 2-21.

Listing 2-21. 21_magic_chart.py

```
# Magic working on Charts
import matplotlib.pyplot as plt
import numpy as np

s = np.random.logistic(10, 5, size=5)
chart, ax = plt.subplots()
ax.hist(s, bins=15)

# Magic chart
"chart", chart
```

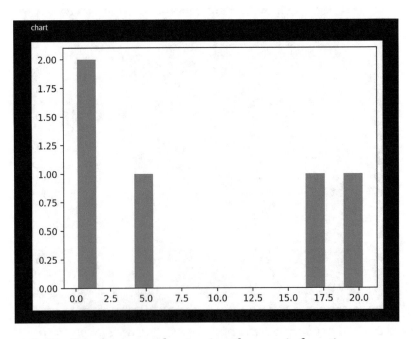

Figure 2-20. *Displaying a chart using the magic function*

Note   The magic feature works in the main Python source app file, as it is currently not supported during file import.

Summary

In this chapter, we learned how text elements can be written in various formats such as for the title, header, subheader, and caption. We also compared different text displayed in our Streamlit application. Then, we looked at text that can be written in formats like plain text, Markdown, LaTeX, and code in different languages.

We saw how data elements can be displayed as tables, metrics, and dataframe objects in the application. Finally, we discussed how the `write()` function can be used to display data elements.

In the next chapter, we will discuss more about visualizing data into charts and graphs with different libraries supported by Streamlit.

CHAPTER 3

Visualization

In the previous chapter, you saw how data in the form of text and tables can be utilized in Streamlit. In this chapter, we will learn what data visualization is and explore Streamlit's functions to visualize data. Later, you'll see how flexible is Streamlit with various Python Libraries to visualize data.

The Importance of Visualization

The visualization of data is important to assist businesses in analyzing recent data trends. When data is represented in the form of charts or graphs, it is easy to decipher and interpret. In data science and machine learning, raw data is visualized to model trends and to reach conclusions about the data.

In addition, visualized data is better for communication between parties and to interrelate the associated factors of the data. The visualized data gives business owners, decision-makers, and stakeholders a better understanding of their data and can help them grow their business.

Visualization in Streamlit

With Streamlit, we can develop customized dashboards for visualizing our data. We can also create any kind of visualization tool in Streamlit as it supports all the Python libraries. We will look at a few main libraries in this

© Sujay Raghavendra 2023
S. Raghavendra, *Beginner's Guide to Streamlit with Python*,
https://doi.org/10.1007/978-1-4842-8983-9_3

chapter such as Plotly, Altair, Seaborn, and Matplotlib. We will also discuss various graphs that can be visualized in the Streamlit application. Now we will discuss the purpose of visualization.

Purpose of Visualization

For data scientists and business analysts, visualization helps to identify patterns and errors in the data provided. For example, during the recent pandemic, scientists visualized data to inform the public about the ongoing situation. Patterns were recognized from the data to identify COVID hotspots. This was done on various types of charts and graphs such as bar charts, line charts, heatmaps, pie graphs, scatter graphs, etc.

Visualizing data helps to solve many issues in business and to develop algorithms accordingly. This in turn speeds up decision-making when solving the inaccuracies identified. In Streamlit, we can create visualization dashboards or tool with less code.

Streamlit Functions

We will now discuss the built-in visualization functions available in Streamlit.

Bar

To represent data points in vertical bars, we can use a bar chart. The built-in `st.bar_chart()` function can plot the data points in a Streamlit application. It is similar to `st.altair_chart()`, which is defined later in the chapter. See Listing 3-1.

Listing 3-1. 01_bar_chart.py

```
import streamlit as st
import pandas as pd
import numpy as np

st.title('Area')

# Defining dataframe with its values
df = pd.DataFrame(
    np.random.randn(40, 4),
    columns=["C1", "C2", "C3", "C4"])

# Bar Chart
st.bar_chart(df)
```

The st.bar_chart() function directly plots a graph by figuring out a column and its associated values, making it easiest to use. In Listing 3-1, we have not defined any values related to the x- and y-axes.

This function is less customizable, and we do not need to specify dataframe values. As shown in Figure 3-1, the values of each column are plotted in a vertical bar with different colors.

Next, we will discuss line charts.

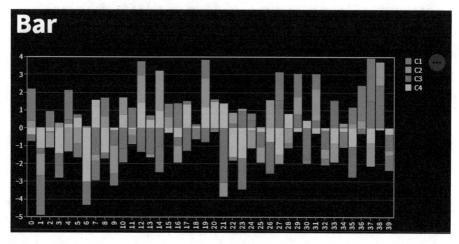

Figure 3-1. *Displaying a bar chart in Streamlit*

Line

Similar to the st.bar_chart() function, the st.line_chart() function automatically detects indices and their values. See Listing 3-2.

Listing 3-2. 02_line_chart.py

```
import streamlit as st
import pandas as pd
import numpy as np

st.title('Area')

# Defining dataframe with its values
df = pd.DataFrame(
    np.random.randn(40, 4),
    columns=["C1", "C2", "C3", "C4"])

# Bar Chart
st.line_chart(df)
```

We have created random values with four columns in Listing 3-2. The line chart is created by passing the dataframe to it.

As shown in Figure 3-2, the values of four columns are represented using four different colors in the chart. The x-axis defines the row number of each column, and the y-axis defines the values in that row.

We will discuss how to implement an area chart in the following section.

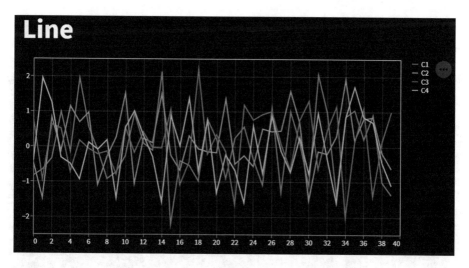

Figure 3-2. *Displaying a line chart in Streamlit*

Area

An area graph or chart is based on the line chart shown in the previous section. The area chart is used to compare different quantitative values of one or more columns. See Listing 3-3.

Listing 3-3. 03_area_chart.py

```
import streamlit as st
import pandas as pd
import numpy as np

st.title('Area')

# Defining dataframe with its values
df = pd.DataFrame(
    np.random.randn(40, 4),
    columns=["C1", "C2", "C3", "C4"])

# Bar Chart
st.area_chart(df)
```

The dataframe is directly given to the `st.area_chart()` function, and the chart looks like Figure 3-3.

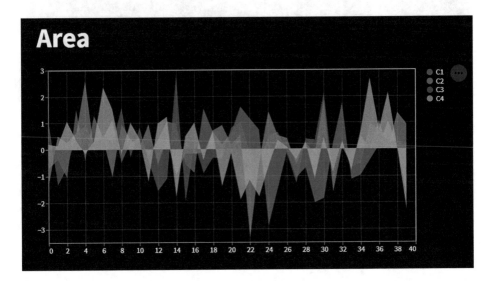

Figure 3-3. *Displaying an area chart in Streamlit*

We will now discuss how to display data points on a specified geographical location on a map.

Note   In some cases, when Streamlit's built-in function for bar, line, and area charts is unable to recognize the values from the data correctly, we can use other charts that are customizable (as shown in the following sections).

Map

To display data points on a map, we will be using the st.map() function
of Streamlit. A scatter plot is applied on top of a map provided by mapbox.
The latitude and longitude specify the position of the map and where the
scatter plot is plotted. See Listing 3-4.

Listing 3-4. 04_map.py

```python
import streamlit as st
import pandas as pd
import numpy as np

st.title('Map')

# Defining Latitude and Longitude

locate_map = pd.DataFrame(
    np.random.randn(50, 2)/[10,10] + [15.4589, 75.0078],
    columns = ['latitude', 'longitude'])

# Map Function
st.map(locate_map)
```

We have generated a few random points that will be displayed on
the map of Dharwad City in India as we have defined the latitude and
longitude (Figure 3-4).

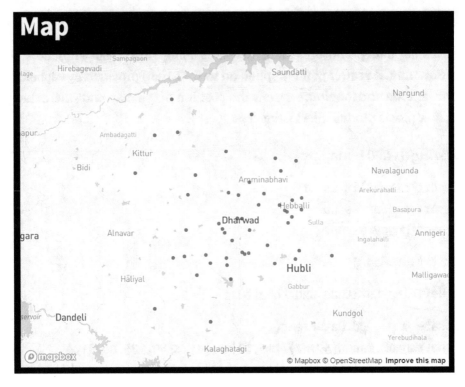

Figure 3-4. *Displaying a map in Streamlit*

The map is positioned at the center. The auto-zoom feature allows us to zoom in and zoom out on the map.

In the following section, we will see how we can represent data in the DOT language.

Graphviz

Graphviz is an open-source Python library in which we can create graphical objects having nodes and edges. The language of nodes and edges is known as the DOT language.

In Listing 3-5, we have implemented a machine learning workflow to train a model.

Listing 3-5. 05_graphviz_dot.py

```python
import streamlit as st
import graphviz as graphviz

st.title('Graphviz')

# Creating graph object
st.graphviz_chart('''
    digraph {
        "Training Data" -> "ML Algorithm"
        "ML Algorithm" -> "Model"
        "Model" -> "Result Forecasting"
        "New Data" -> "Model"
    }
''')
```

The nodes and edges are added in the DOT language, which is easy to understand. We have defined four nodes with connecting edges, as shown in Figure 3-5.

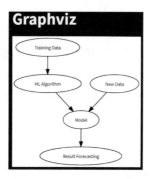

Figure 3-5. *Displaying a Simple ML Model using Graphviz*

We can define the same edge and nodes as shown in Listing 3-6.

Listing 3-6. 06_graphviz.py

```python
import streamlit as st
import graphviz as graphviz

st.title('Graphviz')

# Create a graphlib graph object
graph = graphviz.Digraph()
graph.edge('Training Data', 'ML Algorithm')
graph.edge('ML Algorithm', 'Model')
graph.edge('Model', 'Result Forecasting')
graph.edge('New Data', 'Model')
st.graphviz_chart(graph)
```

We can use this Graphviz code to visualize various ML algorithms or neural networks to help us understand the architecture or flows encountered in them while processing data.

In the following section, we will discuss the Seaborn Python library and how to implement different graphs with it.

Seaborn

Seaborn is a Python visualization library used to get beautifully styled graphs in various color palettes, making graphs more attractive. To get started with Seaborn, we need to install it by using the following command:

```
pip install seaborn
```

Count

The count graph is used to represent a number of occurrences or counts of a categorical variable. Listing 3-7 shows the code for a simple count graph.

Listing 3-7. 07_seaborn_count.py

```
# Import python libraries
import streamlit as st
import seaborn as sns
import pandas as pd
import matplotlib.pyplot as plt

# Data Set
df = pd.read_csv("./files/avocado.csv")

# Defining Count Graph/Plot
fig = plt.figure(figsize=(10, 5))
sns.countplot(x = "year", data = df)
st.pyplot(fig)
```

The categorical data used is year, where the number of occurrences is counted to plot the graph shown in Figure 3-6. The count_plot() function is used from the Seaborn library to display the count graph.

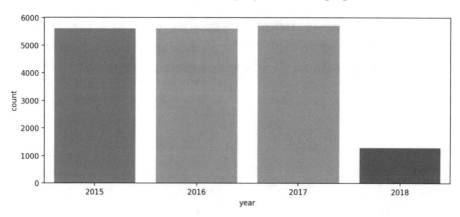

Figure 3-6. *Displaying a Seaborn count chart in Streamlit*

We have used a count plot against the year from our Avocado dataset. Each year is represented in a different color with the number of counts.

Violin

To represent numerical data for more than one group, we can use a violin chart or graph that uses density curves. See Listing 3-8.

Listing 3-8. 08_seaborn_violin.py

```python
# Import python libraries
import streamlit as st
import seaborn as sns
import pandas as pd
import matplotlib.pyplot as plt

# Data Set
df = pd.read_csv("./files/avocado.csv")

# Defining Violin Graph
fig = plt.figure(figsize=(10, 5))
sns.violinplot(x = "year", y="AveragePrice", data = df)
st.pyplot(fig)
```

We have used the violinplot() method to display a violin graph with the x and y parameters defined. The larger the frequency, the larger the width of the curve in the violin region, as depicted in Figure 3-7.

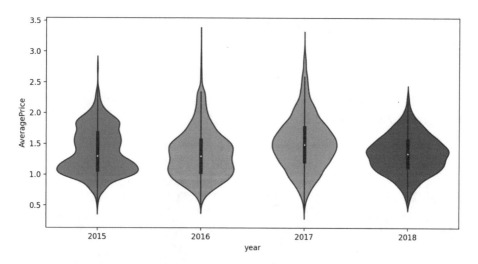

Figure 3-7. *Displaying a Seaborn violin chart in Streamlit*

Next we will check out a strip graph.

Strip

A strip graph represents a summarized univariate dataset. To display all the observations, we can use a strip graph. See Listing 3-9.

Listing 3-9. 09_seaborn_strip.py

```
# Import python libraries
import streamlit as st
import seaborn as sns
import pandas as pd
import matplotlib.pyplot as plt

# Data Set
df = pd.read_csv("./files/avocado.csv")

# Defining Strip Plot
```

```
fig = plt.figure(figsize=(10, 5))
sns.stripplot(x = "year", y="AveragePrice", data = df)
st.pyplot(fig)
```

As shown in Figure 3-8, we can use a strip plot as an alternative to a histogram or density graph.

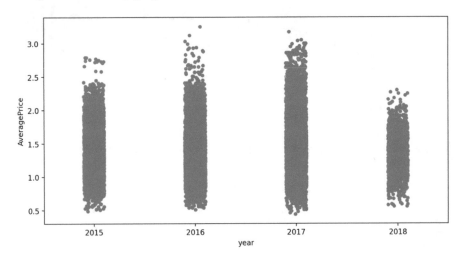

Figure 3-8. *Displaying a Seaborn strip chart in Streamlit*

Altair

An altair is one of the most used statistical visualization python library based on Vega. The visualization process is handled by the Altair library once we define values for x and y. We can also limit the size and set the color of the graphs that are displayed in the application. At first we need to install the Altair library. To install it, we will use the following command:

```
pip install altair
```

Altair supports all type of charts/graphs, but we will discuss only a few types of graphs in Altair.

Boxplot

We can use a boxplot to represent data in terms of skewness by showing data quartiles and averages. See Listing 3-10.

Listing 3-10. 10_altair_boxplot.py

```
import altair as alt
import streamlit as st
import pandas as pd

#Read albany Dataset
df = pd.read_csv("./files/albany.csv")

# Box Plot
box_plot = alt.Chart(df).mark_boxplot().encode(
    x = "Date",
    y = "Large Bags"
)

st.altair_chart(box_plot)
```

In the previous code, we used the mark_boxplot() method to display boxplots. Later we used the st.altair_chart() function to display the graph in our application. The boxplot is divided into five different parts, as displayed in Figure 3-9.

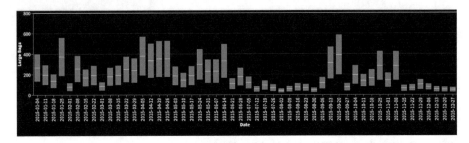

Figure 3-9. *Displaying an Altair box plot chart in Streamlit*

The white band in the middle represents the median.

Next, we will look at how to create and display an area chart in our application.

Area

As we know, an area graph or plot is displayed to represent quantitative data based on line charts. See Listing 3-11.

Listing 3-11. 11_altair_area.py

```
import altair as alt
import streamlit as st
import pandas as pd

#Read albany Dataset
df = pd.read_csv("./files/albany.csv")

# Area Plot
area = alt.Chart(df).mark_area(color="orange").encode(
    x = "Date",
    y = "Large Bags"
)

st.altair_chart(area)
```

We have used the `mark_area()` method to create the area graph. We can see that it is similar to the built-in function that we discussed earlier. See Figure 3-10.

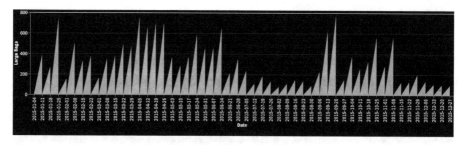

Figure 3-10. *Displaying an Altair area chart in Streamlit*

The additional shading on the line chart distinguishes an area chart. Finally, we will discuss visualizing data into a heatmap.

Heatmap

A heatmap is graphical representation of values in color. The values determine the intensity of color corresponding to two dimensions. See Listing 3-12.

Listing 3-12. 12_altair_heatmap.py

```
import altair as alt
import streamlit as st
import pandas as pd

#Read albany Dataset
df = pd.read_csv("./files/albany.csv")

heat_map  = alt.Chart(df).mark_rect().encode(
        alt.Y('AveragePrice:Q'),
        alt.X('Large Bags:Q'),
        alt.Color('AveragePrice:Q'),
```

```
        tooltip = ['AveragePrice', 'type', 'Large
Bags', 'Date']
    ).interactive()
st.altair_chart(heat_map)
```

The mark_rect() function from the Altair library is used to create heatmaps. We have defined Large Bags and AveragePrice as parameters for the x-axis and y-axis to which the heatmap is produced.

You can see the color intensity range at the right corner of Figure 3-11.

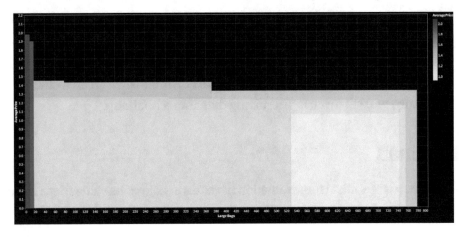

Figure 3-11. *Displaying an Altair heatmap in Streamlit*

We can conclude that an increase in value will increase the color intensity in the graph.

In the next section, we will discuss one of the most widely used interactive Python libraries, called Plotly. We will use Plotly to display graphs in our Streamlit application.

Note   The mark() method is used to represent the type of graph (like line, area, heatmap, etc.) that will be created.

Plotly

Plotly is a Python library that can make high-quality interactive graphs. It supports different types of graphs such as bar charts, line charts, boxplots, pie charts, scatter plots, histograms, etc. We will see how this Python library can be used in our Streamlit application.

First, we need to install the Plotly library by using the `pip` command.

```
pip install plotly
```

Now we will see various graphs that can be implemented in Streamlit using Plotly. We will be using the Avocado dataset for visualizing graphs.

Pie

For the Avocado dataset, we will be visualizing a percentage of organic and conventional avocados.

In Listing 3-13, the first function, `figure()`, is a blank page that is used to plot our pie chart within it. The `Pie()` method creates a pie chart with labels and values as input parameters.

Listing 3-13. 13_plotly_pie.py

```
import streamlit as st
import plotly.graph_objects as go
import pandas as pd

#Read Avocado Dataset
data = pd.read_csv("./files/avocado.csv")

st.header("Pie Chart")

# Implementing Pie Plot
pie_chart = go.Figure(
    go.Pie(
```

```
    labels = data.type,
    values = data.AveragePrice,
    hoverinfo = "label+percent",
    textinfo = "value+percent"
))
```

```
st.plotly_chart(pie_chart)
```

There are two more optional parameters named hoverinfo and textinfo where information is displayed when the mouse is hovered on the pie chart. The function plotly_chart allows us to add a defined pie chart in our application, as shown in Figure 3-12.

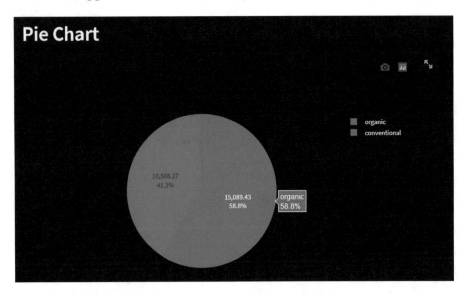

Figure 3-12. *Displaying a Plotly pie chart in Streamlit*

Our pie chart is divided into two parts: organic and conventional. When the mouse is hovered over a part, we can see the name and percentage of that particular section of pie chart. The percentage and total number of avocados corresponding to organic and conventional are displayed in different color. Next, we will move to a donut chart, which is similar to a pie chart.

Donut

Plotly has created a different submodule to plot donut charts. The name of the module is express(). We need to import the express() module in our code. See Listing 3-14.

Listing 3-14. 14_plotly_donut.py

```
import streamlit as st
import pandas as pd
import plotly.express as px

#Read Avocado Dataset
data = pd.read_csv("./files/avocado.csv")

st.header("Donut Chart")

# Donut Chart
donut_chart = px.pie(
    names = data.type,
    values = data.AveragePrice,
    hole=0.25,
)

st.plotly_chart(donut_chart)
```

We have provided same dataset values that are used for pie charts in our earlier example. The pie() method in the Express module takes three parameters: names, values, and hole. The hole specifies the radius of the hole in a donut. See Figure 3-13.

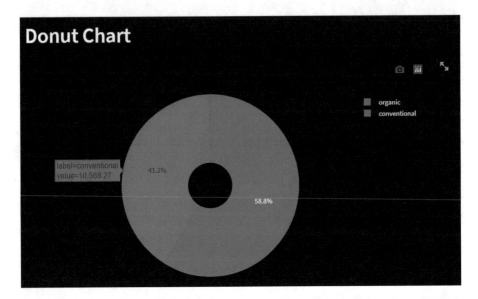

Figure 3-13. *Displaying a Plotly donut chart*

Note   We can convert a donut into a pie chart without specifying a hole parameter.

In the next section, we will discuss how to implement a scatter graph in our application.

Scatter

We will now implement a scatter plot in our application by using the scatter() method from the Plotly library. See Listing 3-15.

Listing 3-15. 15_plotly_scatter.py

```python
import streamlit as st
import pandas as pd
import plotly.express as px

#Data Set
data = pd.read_csv("./files/avocado.csv")

st.header("Scatter Chart")

#Scatter
scat = px.scatter(
    x = data.Date,
    y = data.AveragePrice
)

st.plotly_chart(scat)
```

The x- and y-axes in the scatter() function are the parameters in list form. See Figure 3-14.

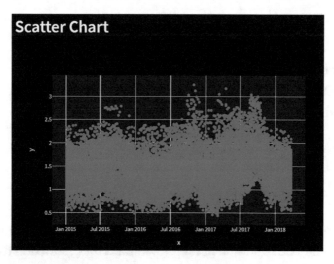

Figure 3-14. Displaying a Plotly scatter chart

The scatter graph or chart represents values in the form of dots. Each dot is associated with the x and y values.

Line

We can create a line graph using the line() method from the Express module in Plotly. The x- and y-axes are the parameters used in a line() function. See Listing 3-16.

Listing 3-16. 16_plotly_line.py

```
import streamlit as st
import pandas as pd
import plotly.express as px

# Data Set
data = pd.read_csv("./files/avocado.csv")

# Minimizing Dataset
albany_df = data[data['region']=="Albany"]
al_df = albany_df[albany_df["year"]==2015]

#Line
line_chart = px.line(
    x = al_df["Date"],
    y = al_df["Large Bags"]
)

st.header("Line Chart")

st.plotly_chart(line_chart)
```

We have minimized the data with respect to the Albany region. The title parameter is an optional parameter in the line() function that provides the title to the line chart of our choice. See Figure 3-15.

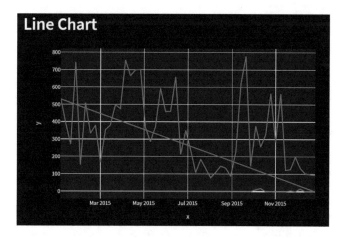

Figure 3-15. *Displaying a Plotly line chart*

We can also change properties such as the color by defining the update_traces() function in our code.

```
line_chart.update_traces(line_color = "orange")
```

In the next section, we will look at a bar graph and its associated subplots implementation in the application.

Bar

We can represent a bar graph using the express.bar() function. Similar to a line chart, a bar graph also has x and y parameters with an optional title parameter. See Listing 3-17.

Listing 3-17. 17_plotly_bar.py

```
import streamlit as st
import pandas as pd
import plotly.express as px

# Data Set
```

```
data = pd.read_csv("./files/avocado.csv")

# Minimizing Dataset
albany_df = data[data['region']=="Albany"]
al_df = albany_df[albany_df["year"]==2015]

# Bar graph
bar_graph = px.bar(
    al_df,
    title = "Bar Graph",
    x = "Date",
    y = "Large Bags"
    )

st.plotly_chart(bar_graph)
```

In the bar() function, we have defined the dataset name and given column names directly to x and y parameters as values, as shown in Figure 3-16.

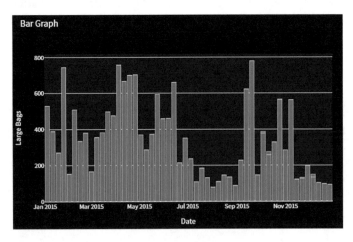

Figure 3-16. *Displaying a Plotly bar graph*

We can also change the color by introducing the color parameter in the bar() function, as shown in Listing 3-18; the results are shown in Figure 3-17.

Listing 3-18. 18_plotly_bar_color.py

```
#Bar Color
bar_graph = px.bar(
    x = al_df["Date"],
    y = al_df["Large Bags"],
    title = "Bar Graph",
    color=al_df["Large Bags"]
)
```

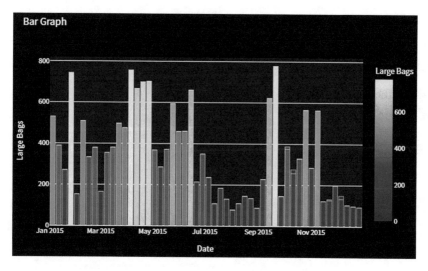

Figure 3-17. *Displaying a Plotly bar chart with color intensity*

The color intensity is associated with the value defined in it. The color intensity in the bar graph helps to understand the range of values depicted.

Bar Horizontal

We can flip the bars in a bar graph to be horizontal. To flip a bar graph to be horizontal, we need to specify the orientation parameter in the bar() function. See Listing 3-19 and the output in Figure 3-18.

Listing 3-19. 19_plotly_bar_horizontal.py

```python
import streamlit as st
import pandas as pd
import plotly.express as px

# Data Set
data = pd.read_csv("./files/avocado.csv")

# Minimizing Dataset
albany_df = data[data['region']=="Albany"]
al_df = albany_df[albany_df["year"]==2015]

# Horizontal Bar Graph
bar_graph = px.bar(
    al_df,
    x = "Large Bags",
    y = "Date",
    title = "Bar Graph",
    color="Large Bags",
    orientation='h'
)
```

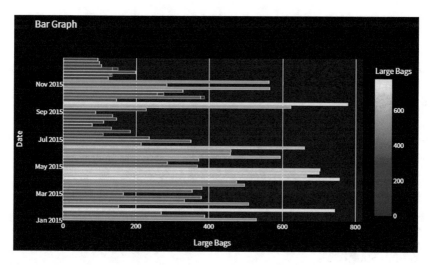

Figure 3-18. *Displaying a Plotly horizontal bar graph*

Finally, we will see how to implement subplots using Plotly in a Streamlit application.

Subplots

Subplots are used to load different graphs next to each other. In Plotly, we can use the make_subplots() method to create multiple graphs to be placed next to one other or to be stacked one below the other. To define subplots, we need to import the potly.subplots module. See Listing 3-20.

Listing 3-20. 20_plotly_subplots.py

```
import streamlit as st
import pandas as pd
import plotly.graph_objects as go
from plotly.subplots import make_subplots

# Data Set
data = pd.read_csv("./files/avocado.csv")
```

```
# Minimizing Dataset
albany_df = data[data['region']=="Albany"]
al_df = albany_df[albany_df["year"]==2015]

fig = make_subplots(rows=3, cols=1)

# First Subplot
fig.add_trace(go.Scatter(
    x=al_df["Date"],
    y=al_df["Total Bags"],
), row=1, col=1)

# Second SubPlot
fig.add_trace(go.Scatter(
    x=al_df["Date"],
    y=al_df["Small Bags"],
), row=2, col=1)

# Third SubPlot
fig.add_trace(go.Scatter(
    x=al_df["Date"],
    y=al_df["Large Bags"],
), row=3, col=1)

st.plotly_chart(fig)
```

In Listing 3-20, we create stacked graphs one below another by defining three rows and a single column. To place graphs next to each other, we can increase the number of columns. See Figure 3-19.

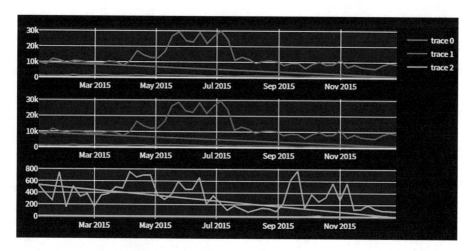

Figure 3-19. *Displaying Plotly subplots*

Subplots can contain any type of graphs. In our example, we have used line graphs in all three rows.

Note   Graphs can be stacked one below the other or placed next to one another by defining the columns in Streamlit. The columns act as container in which graphs are placed.

Summary

Visualization is an important aspect in data science and machine learning. The visualizing techniques help data scientists understand the data better. In this chapter, we started by understanding visualization basics, and then we discussed how we display data points in our application using Streamlit's built-in functions.

Finally, we discussed and implemented different visualizing techniques that are available in Python for data science and machine learning developers.

In the next chapter, we will discuss how to implement media elements in an application.

CHAPTER 4

Data and Media Elements

In this chapter, we will focus on data that we can see and hear. This type of data is in the form of images, audio, videos, or animations. This is known as *multimedia data*. An application may contain any one or a combination of these types of data.

As machine learning models can be developed based on images, audio, or video, we will see how we can display this multimedia data in our application so that we can use these techniques in our final chapters of project building. Now let's get started with images where most of the machine learning models are developed around object detection, object recognition, and image segmentation.

Images

One of the major multimedia files that is needed in an ML/DS application is an image. An image is used generally to visualize the results related to the machine learning models. In this section we will explore images so we can build efficient image applications in the future.

In our example, we have stored images in a directory named `files` that we will be importing to our application. This helps to separate the images from the code. See Listing 4-1.

© Sujay Raghavendra 2023
S. Raghavendra, *Beginner's Guide to Streamlit with Python*,
https://doi.org/10.1007/978-1-4842-8983-9_4

Listing 4-1. 01_image_local.py

```
import streamlit as st

st.write("Displaying an Images")

# Displaying Image by specifying path
st.image("files/animal7.jpg")

#Image Courtesy by unplash
st.write("Image Courtesy: unplash.com")
```

Figure 4-1 shows the results of this code.

Figure 4-1. *Displaying a JPG image in an application*

We can also use any Python library such as OpenCV, PIL, etc., that supports opening an image. It is not compulsory to use any one library. Later, we need to pass the same image file stored in a variable in the image() function of Streamlit.

Listing 4-2 and Figure 4-2 show how an image can be displayed using a URL.

Listing 4-2. 02_image_url.py

```
import streamlit as st

st.write("Displaying an Images")

# Displaying Image URL
st.image("https://tinyurl.com/322vu3ab")

# Image Courtesy
st.write("Courtesy: unplash.com")
```

Figure 4-2. *Displaying an image with a URL*

The syntax for the image() function with other parameters is as follows:

st.image(image, caption, width, use_column_width, clamp, channels, output_format)

While displaying an image in an application, there are various parameters associated with it, as described in Table 4-1.

Table 4-1. *Parameters of the image() Function*

Parameter	Description
image (numpy.ndarray, [numpy.ndarray], str, [str], or BytesIO)	This specifies the image that will be displayed in the application. An image can be monochrome and have a shape (w,h) or (w,h,1) or a color image with the shape (w,h,3) or an RGBA image with (w,h,4). An image can be a URL or path from local storage or SVG XML string or list of more than one image. (Here, (w,h) means the height and width of an image.)
caption (str or string list)	This provides information about an image that is displayed.
width (int or None)	This will help to fix the width of an image that will be displayed. It is set when SVG images are used as SVG images are huge in size. The width can also be adjusted when multiple images are displayed using the column width.
use_column_width ('auto' or 'always' or 'never')	When we set this to auto, the exact size of an image is displayed without exceeding the column width. When we set the column width as "always" or True, then the image width is set to the column width, and when we set it to "never" or False, then the image width is set to its normal size. This parameter when set overwrites the width parameter.

(continued)

Table 4-1. (*continued*)

Parameter	Description
clamp (bool)	The range of clamp values to be selected for an image is from 0 to 255 per channel. This value can be used when the image is in byte array format and skipped when images are defined using a URL. When the clamp value is not set and an image has a value out of range, then an error is raised.
channels ('RGB' or 'BGR')	The channel represents the color information of an image denoting the format used when the image is an nd.array. The channel is by default set to RGB format. The values associated to each channel are as follows: [:,:,0] -> red channel [:,:,1] -> green channel [:,:,2] -> blue channel When we are reading images from Python libraries like OpenCV, we need to set the channel to BGR instead of RGB.
Output_format ('JPEG', 'PNG', or 'auto')	This parameter defines the format to be used when we are transferring the image data. Photos can use JPEG format with lossy compression, while diagrams can use PNG format for lossless compression. When we set the value to "auto," then the compression type is set based on the type and format of an image argument.

One more important feature is that computer vision techniques can be applied for image manipulations using various Python libraries supported by Streamlit.

Next we will see how we can display multiple images in an application.

Note   When we are displaying from a URL or local storage, the image displayed in our application will have the same dimensions (i.e., height and width) of that image. We can resize an image using any Python library before calling into the `st.image` function.

Multiple Images

We can display more than one image in our application by listing the images in the `image()` function. This section shows how we can allow multiple images.

In Listing 4-3, we are defining the width that will be assigned to each image displaying in our application in a column and grid structure. We can adjust the width as we see fit and will be learning more about columns and the grid structure of the Streamlit application in Chapter 7.

Listing 4-3. 03_multiple_local.py

```
import streamlit as st

# Image Courtesy
st.write("Diplaying Multiple Images")

# Listing out animal images
animal_images = [
    'files/animal1.jpg',
    'files/animal2.jpg',
    'files/animal3.jpg',
    'files/animal4.jpg',
```

```
    'files/animal5.jpg',      'files/animal6.jpg',
    'files/animal7.jpg',
    'files/animal8.jpg',
    'files/animal9.jpg',
    'files/animal10.jpg'

]


# Displaying Multiple images with width 150
st.image(animal_images, width=150)

# Image Courtesy
st.write("Image Courtesy: Unplash")
```

Figure 4-3 shows the results.

Figure 4-3. *Displaying multiple images*

Note   An image can be called using a URL or local directory.

We have seen how multiple images are defined in our application and can be used in recommendation systems for book, movies, etc.

In the next section, we will display an image as a background.

Background Image

We can set an image as a background image of an application by embedding it in Markdown language. Let's see how an image is displayed; see Listing 4-4.

Listing 4-4. 04_background_image.py

```python
import streamlit as st
import base64

# Function to set Image as Background
def add_local_backgound_image_(image):
    with open(image, "rb") as image:
        encoded_string = base64.b64encode(image.read())

    st.write("Image Courtesy: unplash")

    st.markdown(
    f"""
    <style>
    .stApp {{
        background-image: url(data:files/
{"jpg"};base64,{encoded_string.decode()});
        background-size: cover
    }}
    </style>
    """,
    unsafe_allow_html=True
    )

st.write("Background Image")

# Calling Image in function
add_local_backgound_image_('files/animal7.jpg')
```

First, we open an image and convert it to base64-encoded format. Then, we define a function wherein an image is called in the background using Markdown. We can define any image format for a background image. Figure 4-4 illustrates the results.

Figure 4-4. *Image as background*

In the next section, we will see how we can resize our image to be displayed in our application.

Resizing an Image

In machine learning, images are trained at a uniform level to extract features from them and obtain an image model. When models are deployed and we want to test our model with new images, then the images need to be resized to the size of the trained model. Hence, we will see how the resizing of an image is done. The same will be applied while image uploads are used. See Listing 4-5 and Figure 4-5.

Listing 4-5. 05_image_resize.py

```
import streamlit as st
from PIL import Image
```

```
original_image = Image.open("files/animal9.jpg")

# Display Original Image
st.title("Original Image")
st.image(original_image)

# Resizing Image to 600*400
resized_image = original_image.resize((600, 400))

#Displaying Resized Image
st.title("Resized Image")
st.image(resized_image)
```

Figure 4-5. *Displaying a) the original image and b) the resized image*

Note   An image is resized to match the trained model images.

To resize an image, we used the PIL Python library. We can use any Python library that supports image resizing in our application.

In the next section, we will look at audio data.

Audio

Audio is a type of multimedia file where we can hear only the sound from the file. We can use an audio file in our application by using the st.audio() function, as shown in Listing 4-6.

We need to open our audio file by passing the audio file name in binary mode using the rb parameter in our first statement. Later, we use the audio() function from Streamlit to display the audio file in our application by providing the saved variables for the audio as input. See Figure 4-6.

Listing 4-6. 06_audio.py

```
import streamlit as st

# Open Audio using filepath with filename
sample_audio = open("files/audio.wav", "rb")

#Reading Audio File
audio_bytes = sample_audio.read()

# Display Audio using st.audio() function with start time
set to 20
st.audio(sample_audio, start_time = 20)

# Printing Audio Courtesy
st.write("Audio Courtesy: https://file-examples.com/index.php/
sample-audio-files/sample-wav-download/")
```

Figure 4-6. *Displaying an audio file in an application*

The control buttons display with the audio files in our application. The type of audio file supported is the MIME type. We can set the start time of the audio file from which it will start playing. The start time is always an integer value.

We can also use a URL in place of an audio file that is stored locally on disk. This is one of best features as we can directly access the audio file stored in the cloud or on an online server.

Listing 4-7 shows an example of an audio file being accessed by using a URL. See Figure 4-7.

Listing 4-7. 07_audio_url.py

```python
import streamlit as st

# Open Audio using filepath with filename and read the
audio file
sample_url = st.audio("https://www.learningcontainer.com/wp-
content/uploads/2020/02/Kalimba.mp3")

st.write("Audio Courtesy: https://www.learningcontainer.com/
sample-audio-file/")
```

Figure 4-7. *Displaying an audio file using a URL*

Streamlit supports most of the audio formats like MP3, WAV, OGG, etc. When no audio format is specified, then Streamlit will automatically take audio in audio/wav format. In our next section, we will look into displaying video files in our Streamlit application.

Video

We can upload any video to our application using the st.video() function. Similar to audio files, we can specify the starting time of the video uploaded instead of the default time starting from 0. Here, the integer value specifies the time in seconds. Let's use the function to view video in our application. See Listing 4-8.

Listing 4-8. 08_video_local.py

```
import streamlit as st

# Open Video using filepath with filename and read the
video file
sample_video = open("files/ocean.mp4", "rb").read()

# Display Video using st.video() function
st.video(sample_video, start_time = 10)

st.write("Video Courtesy: https://filesamples.com/formats/mp4")
```

In our application, we have provided a start time as 10, which means the video will play from the 10th second. The video won't start unless we click the play button. See Figure 4-8.

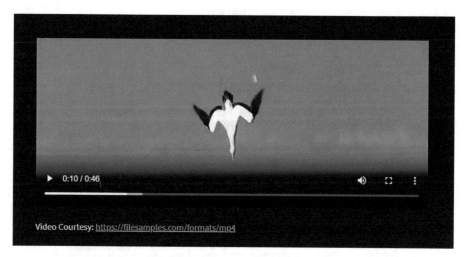

Figure 4-8. *Displaying a video file in our application*

The controls such as play, pause, full screen, and volume are displayed. The st.video also provides features such as playback speed, download ability, and picture-in-picture that can be explored by clicking the button with three vertical dots.

In some cases, videos encoded in MP4V format are not supported in standard browsers but can be used as an export option by importing an OpenCV library in Python. We can also display the MP4V-encoded video by converting to H.264 format, which is supported in Streamlit.

We can also upload a video in our application using a URL. In Listing 4-9, we will be using a YouTube video link in our application. See Figure 4-9.

Listing 4-9. 09_video_youtube.py

```
import streamlit as st

# Displaying Video using youtube URL
st.video("https://www.youtube.com/watch?v=OMkEVX23BdM")

# Courtesy by youtube channel
st.write("Video Courtesy: National Geographic Channel")
```

Figure 4-9. *Displaying video using a YouTube URL*

We can also provide a start time to the YouTube video along with the URL. The size of the video is the default set by Streamlit and cannot be changed. To change the height and width of the video, we need to embed the video in the HTML code and then run it in a Streamlit application.

Note   When we use a URL to display video in our application, there is no need to read the video file as in the case of a local upload.

Balloon

Listing 4-10 creates an animation of balloons flying in our application. See Figure 4-10.

Listing 4-10. 10_balloons.py

```
import streamlit as st

# Animated Snowflakes
st.balloons()
```

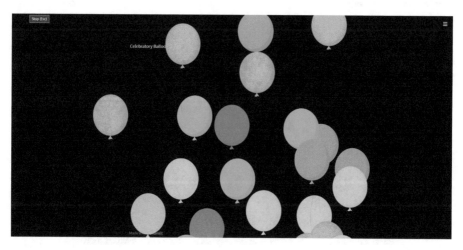

Figure 4-10. *Displaying animated balloons moving upward*

The animated balloons start from the bottom of our application and move toward the top. The animation is very fast in nature, and we can add the time.sleep() function to delay the animation's start. We will look into one more built-in animation function available in our next section.

Snowflake

The animated snowflake was introduced when the Snowflake company was acquired Streamlit. Therefore, this animation is available in Streamlit version 1.17 and above. The function for the animated snowflake is snow( ). Listing 4-11 shows the code, and Figure 4-11 shows the output.

Listing 4-11. 11_snowflake.py

```
import streamlit as st

# Animated Snowflakes
st.snow()
```

Figure 4-11. *Displaying animated snowflakes*

Note   The built-in snowflake function works only in Streamlit version 1.17 and above.

The animated snowflakes behave like they are falling from the sky; i.e., they starts animating from top to bottom in an application.

Emojis

Emojis are mini icons or images that are generally used to describe expressions, feelings, or messages in digital communication and now in web applications. This was possible when the Internet started to flourish and Unicode became acceptable. These emojis are supported by Streamlit and hence can be used in our application.

There are two ways to add an emoji: one is to add an emoji using an ASCII value, and the other is to add the shortcode for the respective emoji. You can decide which one you prefer based on the following example. See Listing 4-12 and Figure 4-12.

Listing 4-12. 12_emojis.py

```
import streamlit as st

# Emojis with/without shortcodes
emojis = """:rain_cloud: :coffee: :love_hotel:  👫👫
:couple_with_heart: 💍"""

# Displaying Shortcodes
st.title(emojis)
```

Figure 4-12. *Displaying emojis*

Figure 4-12 shows the emojis we have displayed with and without shortcodes. Here, shortcodes like `:rain_cloud:`, `:coffee:`, `:love_hotel:`, and `:couple_with_heart:` are displayed. We can also write emojis in text format, as we have written for the male and female superheroes. It is preferred to write emojis using shortcodes as we know what the picture is depicting in an application.

You can check for all emoji shortcodes that are available for Streamlit application at `https://streamlit-emoji-shortcodes-streamlit-app-gwckff.streamlitapp.com/`. You can also display any other emoji that is not available using other emoji sites.

We can use emojis in place of any of the text covered in Chapter 2.

Summary

In this chapter, we saw multimedia data in various formats such as images, audio, and videos to be displayed in our application. We saw how an image can be displayed from local storage or a URL. We also saw how to resize images, define multiple images, and display an image as a background.

We also looked at the balloon and snowflake animations that are available in Streamlit as built-in functions.

Finally, we learned how to display emojis in our data.

In the next chapter, we will look into buttons and sliders and their associated features.

CHAPTER 5

Buttons and Sliders

In this chapter, we will discuss how to create different types of buttons. These buttons are a medium for interaction between the user and the application. Let's start by creating these buttons one by one.

Buttons

A button is created by using the following syntax:

```
st.button('button_name')
```

Let's create a button by using this syntax; see Listing 5-1.

Listing 5-1. 01_button.py

```
import streamlit as st

st.title('Creating a Button')

# Defining a Button
button = st.button('Click Here')

if button:
    st.write('You have clicked the Button')
else:
    st.write('You have not clicked the Button')
```

In Listing 5-1, we have used an `if` condition to figure out whether the button has been clicked or not. If the button is clicked, then the statement "You have clicked the Button" is executed. If the button is not clicked, then the statement "You have not clicked the Button" is active. Figure 5-1 shows the results.

Figure 5-1. *Button output a) before click and b) after click*

Radio Buttons

To create radio buttons, we can use the following syntax:

```
st.radio('button_labels', radio_button_names,
index=index_value)
```

Listing 5-2 creates three radio buttons containing different genders, and Figure 5-2 shows the output.

Listing 5-2. 02_radio_buttons.py

```
import streamlit as st

st.title('Creating Radio Buttons')

# Defining Radio Button
gender = st.radio(
    "Select your Gender",
```

```
    ('Male', 'Female', 'Others'))

if gender == 'Male':
    st.write('You have selected Male.')
elif gender == 'Female':
    st.write("You have selected Female.")
else:
    st.write("You have selected Others.")
```

Figure 5-2. *Radio button*

By default, the first radio button is selected, but you can change it by using indexing. We can use the code in Listing 5-3 to change the index. Figure 5-3 shows the output.

Listing 5-3. 03_radio_buttons_index.py

```
import streamlit as st

st.title('Creating Radio Buttons')

gender = ('Male', 'Female', 'Others')

# Defining Radio Button with index value
gender = st.radio(
```

```
    "Select your Gender",
    gender,
    index = 1)

if gender == 'Male':
    st.write('You have selected Male.')
elif gender == 'Female':
    st.write("You have selected Female.")
else:
    st.write("You have selected Others.")
```

Figure 5-3. *Radio button with defined index selection*

Here, the index value is changed to 1, and therefore the Female radio button is selected. After execution, the output is displayed on the screen.

Check Boxes

To create a check box, we will use the following syntax:

```
st.checkbox('button_label', checkbox_names, index=index_value)
```

Listing 5-4 creates three check boxes with the hobbies Books, Movies, and Sports.

Listing 5-4. 04_checkbox.py

```python
import streamlit as st

st.title('Creating Checkboxes')

st.write('Select your Hobies:')

# Defining Checkboxes
check_1 = st.checkbox('Books')
check_2 = st.checkbox('Movies')
check_3 = st.checkbox('Sports')
```

One or more check boxes can be created as there is no limit on them. Figure 5-4 shows the check boxes before and after selection. We can even select all the available check boxes.

Figure 5-4. *Check box output a) before click and b) after click*

The check box created can be preselected by using the code in Listing 5-5. Figure 5-5 shows the output.

Listing 5-5. 05_checkbox_pre_select.py

```python
import streamlit as st

st.title('Pre-Select')

check = st.checkbox('Accept all Terms and Conditions***',
value=True)
```

Figure 5-5. Preselecting the checkbox output

To preselect a checkbox, we need to add the Boolean value `True` for the value attribute.

Drop-Downs

A drop-down contains options listed within it that are revealed when it is clicked by the user. It is easy to create a drop-down in Streamlit by using the following syntax:

```
st.selectbox(label, options, index)
```

We will create a simple drop-down with multiple options; see Listing 5-6.

Listing 5-6. 06_dropdown.py

```
import streamlit as st

st.title('Creating Dropdown')

# Creating Dropdown
hobby = st.selectbox('Choose your hobby: ',
        ('Books', 'Movies', 'Sports'))
```

Figure 5-6 shows the drop-down that will be displayed in our application.

Figure 5-6. *Displaying a drop-down*

In Figure 5-6, there is no user interaction with the drop-down, so we are not able to see the list of options that we assigned in the code, and by default the first option is selected, similar to the radio box. Figure 5-7 displays all the options provided when we click.

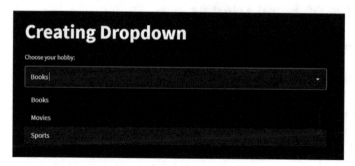

Figure 5-7. *Displaying options in a drop-down*

We can change the default option selected by providing the index value. See Listing 5-7 and Figure 5-8.

Listing 5-7. 07_dropdown_index.py

```python
hobby = st.selectbox('Choose your hobby: ',
        ('Books', 'Movies', 'Sports'),
        index=1)
```

Figure 5-8. *Displaying a drop-down after setting an index value*

Next, we will look into selecting multiple options from the drop-down.

Multiselects

Now we will look at how to multiselect options. See Listing 5-8.

Listing 5-8. 08_mutlti_select.py

```
import streamlit as st

st.title('Multi-Select')

# Defining Multi_Select with Pre-Selection
hobbies = st.multiselect(
        'What are your Hobbies',
        ['Reading', 'Cooking', 'Watching Movies/TV Series',
'Playing', 'Drawing', 'Hiking'],
        ['Reading', 'Playing'])
```

In this multiselect code, we have preselected two hobbies; the output is displayed in Figure 5-9.

Figure 5-9. *Multiselect output*

One can skip the preselection by removing the selected options from the code, as shown in Listing 5-9. See Figure 5-10 and Figure 5-11 for the results.

Listing 5-9. 09_mutlti_select_no_pre_select.py

```python
hobbies = st.multiselect(
    'What are your Hobbies',
    ['Reading', 'Cooking', 'Watching Movies/TV Series',
    'Playing', 'Drawing', 'Hiking']
```

Figure 5-10. *Multiselect output with no preselection*

Figure 5-11. *Multiselect output after selection*

We will now look at how to download a file with a download button.

107

Note   The user may experience higher loading time if multiselection has more than 100 attributes in it. Therefore, it is recommended to use fewer attributes.

Download Buttons

When we allow a user to download results or a file directly from our application, Streamlit provides a download button for the user to click. The file format can be JPG/PNG, TXT, CSV, etc. See Listing 5-10.

Listing 5-10.   10_download_button.py

```
import streamlit as st

st.title("Download Button")

# Creating Download Button
down_btn = st.download_button(
        label="Download Image",
        data=open("./files/fam.jpg", "rb"),
        file_name="lions.jpg",
        mime="image/jpg"
    )
```

In the previous code, we have assigned an image to the download button. When a user clicks the download image button, a JPG file will be downloaded on the local system. See Figure 5-12.

Figure 5-12. *Displaying a download image button a) before clicking and b) after clicking*

Similarly, we can use the download button to download a CSV file. In some cases, we can convert the dataframe analyzing the results to CSV by using a download button. See Listing 5-11.

Listing 5-11. 11_download_dataframe.py

```
st.download_button(
    label="Download CSV",
    data=open("./files/avocado.csv", "rb"),
    file_name='data.csv',
    mime='csv',
)
```

In our example, we have used a CSV file from our local disk and provided to the download button. In the next section, we will see how we can provide a progress bar to the user while they are waiting.

Progress Bars

We have seen progress bars when downloading something from the Web. We can use it in the same way to let the user know that something is still downloading. See Listing 5-12.

Listing 5-12. 12_progress_bar.py

```
import streamlit as st
import time

st.title('Progress Bar')

# Defining Progress Bar
download = st.progress(0)

for percentage in range(100):
    time.sleep(0.1)
    download.progress(percentage+1)

st.write('Download Complete')
```

In Figure 5-13, we see a progress bar that has not started yet, whereas in Figure 5-14 we see a progress bar that has reached 100 percent.

Figure 5-13. *Displaying a progress bar*

Figure 5-14. *Displaying a completed progress bar*

In the next section, we will discuss spinners and when they can be used in our application.

Spinners

A spinner is used to display a message while the user is waiting for some task to execute in our application. Specifically, we can use a spinner while uploading data that is being processed by the application. The message is temporary in nature. See Listing 5-13.

Listing 5-13. 13_spinner.py

```
import streamlit as st
import time

st.title('Spinner')

# Defining Spinner
with st.spinner('Loading...'):
    time.sleep(5)
st.write('Hello Data Scientists')
```

We have given the spinner five seconds of wait time, and it is displayed in Figure 5-15 (a). After the wait time is over, we will display the "Hello Data Scientists" text, as shown in Figure 5-15 (b).

Figure 5-15. *Displaying a spinner a) buffering and b) after the wait time is over*

A spinner is often used when results are being processed in an application.

Summary

In this chapter, we discussed various kinds of buttons that can be used in our applications. We started with a default button and then moved on to radio buttons where we can have multiple options. We also discussed check boxes where we can have multiple options and a choice is made.

We also saw drop-down menus in which a user can choose only one option from the given options. We also saw multiselect menus where we can choose more than one or all the options from the available choices. Then, we discussed using a download button to download files.

Finally, we learned about progress bars and spinners that can be used while processing a task and downloading a file.

In the next chapter, we will discuss more about getting different types of data via a form.

CHAPTER 6

Forms

In this chapter, we will discuss various form attributes used to elicit input from users and provide results accordingly. In ML applications, when we want to test our model or provide results to a user, we need to first collect input from the user. For example, if we want to apply a brightness or contrast value on an image before giving it to a model to predict, user inputs are needed to set these values and analyze the results associated with them. Uploading files also requires user input, and the user/client will expect a result.

When creating a data science or machine learning application, we need to focus on user interactions by making the application dynamic in nature. User interactions are the inputs received from the user and then the results that are displayed. The user inputs can be in the form of text, numbers, date/time selections, and color picker. This chapter will give us in-depth knowledge of the various ways we can get user input.

To get user input from our application, we will be using a text box, text area, date input, number input, time input, and color picker. Let's get started with an input box, which is the first method covered in this chapter.

Text Box

A text input box allows a user to enter data. It is one of the most important elements in a form. In this section let's see how we can create a text box.

© Sujay Raghavendra 2023
S. Raghavendra, *Beginner's Guide to Streamlit with Python*,
https://doi.org/10.1007/978-1-4842-8983-9_6

Listing 6-1 creates a text box that accepts a name as textual data from the user.

Listing 6-1. 01_text_box.py

```
import streamlit as st

st.title("Text Box")

# Creating Text box
name = st.text_input("Enter your Name")

st.write("Your Name is ", name)
```

In Figure 6-1 (a), we can see the input box defined with a name field, whereas Figure 6-1 (b) shows the input data specified by the user, which is shown as the result. To submit the text entered, the user needs to press the Enter key on the keyboard.

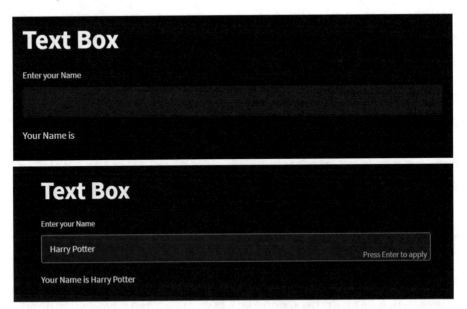

Figure 6-1. a) A text box before entering data and b) the text box after entering data

We can limit the characters to be entered in a text box by using the max_chars parameter. In Listing 6-2, we have limited the characters to a maximum of 10, which is depicted in Figure 6-2 in the corner of the text box.

Listing 6-2. 02_text_box_max_chars.py

```
import streamlit as st

st.title("Text Box")

# Creating Text box with 10 as character limit
name = st.text_input("Enter your Name", max_chars=10)

st.write("Your Name is ", name)
```

Figure 6-2. *Displaying a text box with character limitations*

We can use the same text box as a password box that hides text data because it's a password. To implement a password input box, we will use the code in Listing 6-3. Figure 6-3 shows the result.

Listing 6-3. 03_input_text_password.py

```python
import streamlit as st

st.title("Text Box as Password")

password = st.text_input("Enter your password",
type='password')
```

Figure 6-3. *Displaying a text box as a password field*

We can also implement a character limit on the password input box. We can see the password by clicking the eye icon at the end of the text box, as shown in Figure 6-4.

Figure 6-4. *Displaying a text box as a password field with visible characters*

Now, let's look at a text area where we can get multiline input.

Text Area

A text area is used when we want to get multiple lines of text from the user. This type of format is often used to analyze sentiments that are in the form of comments, reviews, feedbacks, or any opinion given on social media or on a website. We can see how to define a text area in Listing 6-4.

Listing 6-4. 04_text_area.py

```python
import streamlit as st

# Creating Text Area
input_text = st.text_area("Enter your Review")

# Printing entered text
st.write("""You entered:  \n""",input_text)
```

As shown in Figure 6-5, the text area can have multiple lines within this same box. There is also a provision to resize the text area from the application.

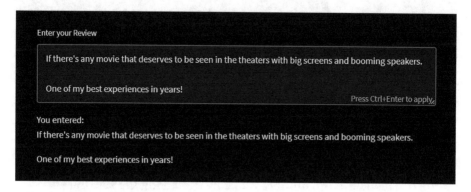

Figure 6-5. *Displaying a text area*

In the next section, we will learn how we can get a number as input in our application.

Number Input

In this input type, a user can enter only numbers in the input box. To define a number as input, we will use the number_input() function from Streamlit. See Listing 6-5 and Figure 6-6.

Listing 6-5. 05_number.py

```
import streamlit as st

# Create number input
st.number_input("Enter your Number")
```

Figure 6-6. *Displaying a number input a) before entering data and b) after entering data*

We can subtract or add numbers by using the - and + provided beside the number box. The numbers can be decimals or integers.

We can also limit the number range by specifying the starting and ending number. We can also set the default value if the user does not specify any value. There is one more parameter in the number_input() function known as step_size. When step_size is set, it will add to the default value provided (i.e., number + step_size) while changing the number from the input box. See Listing 6-6 and Figure 6-7.

Listing 6-6. 06_number_range.py

```
import streamlit as st

# Create number input
num = st.number_input("Enter your Number", 0, 10, 5, 2)

st.write("Min. Value is 0, \n  Max. value is 10")

st.write("Default Value is 5, \n  Step Size value is 2")

st.write("Total value after adding Number entered with step
value is:", num)
```

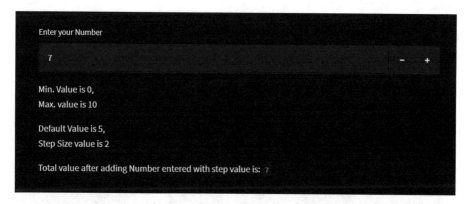

Figure 6-7. *Displaying the number input with all the parameters*

Here in our previous example, we have set the default number as 5 and step_value as 2. The number starts its range from 0 and ends at 10, respectively. When we want to increase or decrease the default value of 5, the number will change by what is in step_value.

Note   The step_value is used to alter a number from the - and + (minus and plus) button provided with the number input box.

In the next section, we will look at how the user can specify the time and date.

Time

We can define the time to represent a specific period to show when the machine learning model was used in our application by the user. It is a 24-hour clock. See Listing 6-7 and Figure 6-8.

Listing 6-7. 07_time.py

```
import streamlit as st

st.title("Time")

# Defining Time Function
st.time_input("Select Your Time")
```

Figure 6-8. *Displaying the current time as the default with drop-down time options*

We cannot set a time before the current time. The time provided by this function is one hour in the future from the current time, with a time gap of 15 minutes.

Note   The current system time is considered the default time by the `time_input()` function.

Next, we will look at the date function given by Streamlit.

Date

The date helps us to define the date with the year, month, and day, respectively. We define a date using the following syntax:

```
st.date_input()
```

We have defined the date function in our application as shown in Listing 6-8.

Listing 6-8. 08_date.py

```
import streamlit as st

st.title("Date")

# Defining Date Function
st.date_input("Select Date")
```

We can also change the default format of the year, month, and date. The date can be picked with a calendar, as shown in Figure 6-9.

Figure 6-9. *Displaying the current date with a calendar*

We can also set on which date the calendar starts and ends by specifying the min_value and max_value values in the date_input() function. See Listing 6-9 and Figure 6-10.

Listing 6-9. 09_date limit.py

```
import streamlit as st
import datetime

st.title("Date")

# Defining Time Function
st.date_input("Select Your Date", value=datetime.date(1989,
12, 25),
    min_value=datetime.date(1987, 1, 1),
    max_value=datetime.date(2005, 12, 1))
```

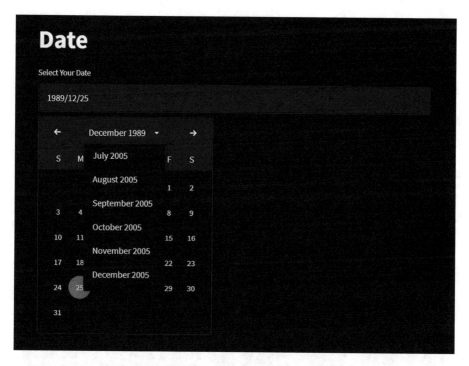

Figure 6-10. *Setting the start and end dates*

We now know how to use the calendar in our application. The next element we will look at is a color selection box.

Note   The default date is set to the current day.

Color

Streamlit provides a unique color selection box as an input feature. By using the color widget, we can select any color of our choice. The color codes are hexadecimal numbers used to define the color. See Listing 6-10 and Figure 6-11.

123

Listing 6-10. 10_color.py

```
import streamlit as st

st.title("Select Color")

# Defining color picker
color_code = st.color_picker("Select your Color")

st.header(color_code)
```

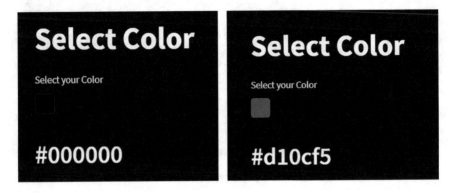

Figure 6-11. *Displaying a color selection widget a) before color selection and b) after color selection*

The default color selected is black, with #000000 as a hex value. The color widget has a slider and also an option to change the RGB value, as shown in Figure 6-12.

Figure 6-12. *Displaying a color selection widget a) with a hex value and b) with an RGB value*

In the next section, we will discuss how files in various formats can be taken as input from the user in our Streamlit application.

File Upload

Sometimes users need to upload files in an ML/DS application for analysis from the local storage. Files might be used for object detection/recognition in an image, forecasting from CSV data, text detection/summarization (NLP technique) from text files, etc.

Uploading a file is one way to input data in an application. The data may be in the form of text file, CSV file, or image. We will see how we can upload such diverse data into our application and also store it in a specific directory. We will begin with text documents.

Text/Docx Document

A file that contains only textual data in it is considered to be a text document. A text file can be created using a text editor like Notepad. We will see how a text document, specifically a .docx file, can be uploaded in our application. To view the contents of a .docx file, we will be using the docx2txt Python library. There are other similar libraries that can be used. See Listing 6-11 and Figure 6-13.

Listing 6-11. 11_new_doc.py

```python
#Import Necessary Libraries
import streamlit as st
import docx2txt

st.title("DOCX & Text Documents")

# Defining File Uploader Function in a variable
text_file = st.file_uploader("Upload Document",
type=["docx","txt"])

# Button to check document details
details = st.button("Check Details")

# Condition to get document details
if details:

    if text_file is not None:

        # Getting Document details like name, type and size
        doc_details = {"file_name":text_file.name, "file_
        type":text_file.type,
                        "file_size":text_file.size}
        st.write(doc_details)
```

```python
# Check for text/plain document type
if text_file.type == "text/plain":
    # Read document as string with utf-8 format
    raw_text = str(text_file.read(),"utf-8")
    st.write(raw_text)

else:
    # Read docx document type
    docx_text = docx2txt.process(text_file)
    st.write(docx_text)
```

Figure 6-13. *Displaying a document upload section, before upload*

As shown in Figure 6-14, there is a browse button that enables the user to upload a file from the local storage or simply drag and drop the file into the upload area.

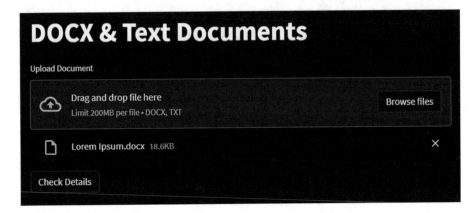

Figure 6-14. *Displaying a document upload section, after upload*

The size limit specified by Streamlit is 200 MB and cannot be uploaded beyond it. The document type to be uploaded can also be seen in the application.

We will be displaying file details such as the name of the uploaded file, its format type, and its size. These details after the user clicks the Check Details button. We will see all the textual data in it, as shown in Figure 6-15.

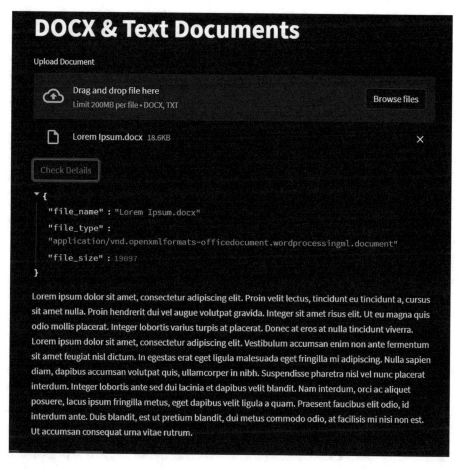

Figure 6-15. *Displaying the contents of the .docx file that was uploaded*

Similar to the `.docx` file, clicking the Check Details button will display the contents of the text document in our application, as shown in Figure 6-16.

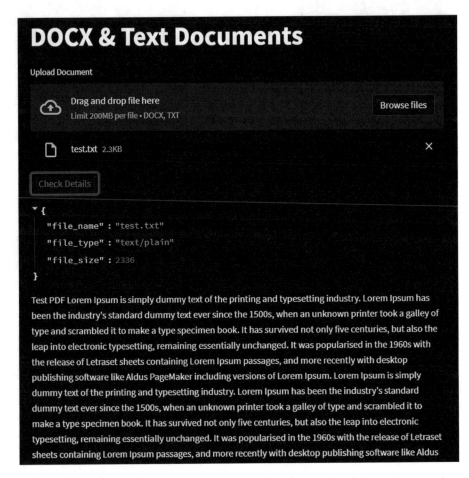

Figure 6-16. Displaying the contents of the text file that was uploaded

---

Note   The data in the .docx file is *lorem ipsum* text that is used as dummy data.

PDF Upload

PDF is one of the most commonly used data types. It contains textual and image data, which makes it different from the plain .docx file shown earlier. We will upload the file and display the content in it. See Listing 6-12.

Listing 6-12. 12_pdf.py

```python
import streamlit as st
import pdfplumber

st.title("PDF File")
pdf_file = st.file_uploader("Upload PDF", type=["pdf"])

details = st.button("Check Details")
if details :

    if pdf_file is not None:

        pdf_details = {"filename":pdf_file.name,
        "filetype":pdf_file.type,
                        "filesize":pdf_file.size}
        st.write(pdf_details)
        pdf = pdfplumber.open(pdf_file)
        pages = pdf.pages[0]
        st.write(pages.extract_text())
    else:
        st.write("No PDF File is Uploaded")
```

We have installed the pdfplumber Python library using the pip command (pip install pdfplumber) in the environment defined for Streamlit.

As shown in Figure 6-17, we are displaying the contents and metadata of a PDF file by hitting the Click Details button. The metadata consists of the file_name, file_type, and file_size values of the uploaded PDF.

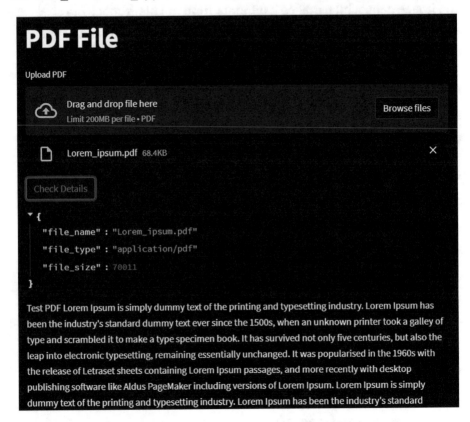

Figure 6-17. *Displaying the contents of the PDF file that was uploaded*

In the next section, we will discuss the tabular data format that can be uploaded in our application.

Dataset Upload

Up until now we have seen files containing only textual data that are uploaded in our application. We will now discuss data that is available in tabular format and that is the CSV type. CSV documents contain textual data separated by comma, making a table- or column-like structure. We will see how to upload such files and visualize the data in tables. See Listing 6-13.

Listing 6-13. 13_csv.py

```python
import streamlit as st
import pandas as pd

st.title("CSV Data")
data_file = st.file_uploader("Upload CSV",type=["csv"])

details = st.button("Check Details")
if details:

    if data_file is not None:

        file_details = {"file_name":data_file.name, "file_
        type":data_file.type,
                        "file_size":data_file.size}

        st.write(file_details)
        df = pd.read_csv(data_file)
        st.dataframe(df)
    else:
        st.write("No CSV File is Uploaded")
```

We have used the Pandas library to display the results in table format for the uploaded CSV file. You can download this data from Kaggle (https://www.kaggle.com/datasets/neuromusic/avocado-prices). We have defined only the CSV type to be uploaded. See Figure 6-18.

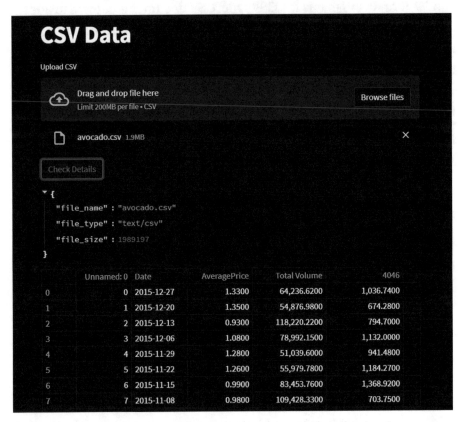

Figure 6-18. *Displaying the contents of a CSV file that was uploaded*

We can see the contents and complete details of the CSV file uploaded by clicking the Check Details button.

So far, we have discussed how to upload various file types in an application. Next, we will see how we can upload images.

Image Upload

In this section, we will discuss how to upload an image in our application. Unlike other document files, images can have different file formats like JPG, PNG, etc., which is read after upload. We need to specify the file format that will be accepted as input data in our application. Listing 6-14 shows how image upload can be done.

Listing 6-14. 14_image_upload.py

```python
import streamlit as st
from PIL import Image
import io

st.title("Upload Image")
image_file = st.file_uploader("Upload Images",
type=["png","jpg","jpeg"])

check_details = st.button("Check Details")
if check_details:

    if image_file is not None:

            # To See details
            image_details = {"file_name":image_file.name,
            "file_type":image_file.type,
                            "file_size":image_file.size}
            st.write(image_details)

            # To View Uploaded Image
            image_data = image_file.read()
            image = Image.open(io.BytesIO(image_data))
            st.image(image, width=250)

    else:
        st.write("No Image File is Uploaded")
```

In the example, we have specified image file formats that can be uploaded in the type parameter of the st.file_uploader() function. The check_details button will provide the metadata of the file along with displaying the uploaded image, as shown in Figure 6-19.

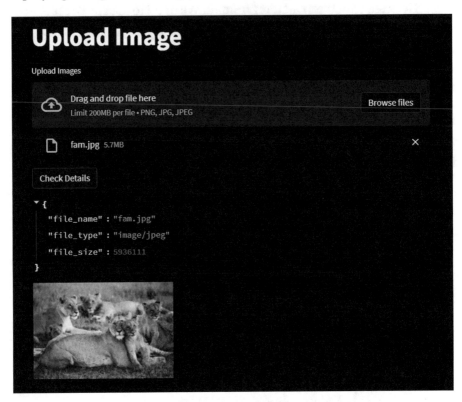

Figure 6-19. *Displaying a single image file that was uploaded*

Uploading Multiple Images

Earlier we saw how to upload an image as a single file, which is often used in ML applications to get results from the model developed. We can also upload multiple images at the same time to get results or train a model. First, we will learn how we can get multiple image files instead of a single default file. See Listing 6-15.

Listing 6-15. 15_image_multiple.py

```
import streamlit as st
import io
from PIL import Image

uploaded_files = st.file_uploader("Multiple Image Uploader",
type=['jpg','jpeg','png'],
                    help="Upload Images in jpg, jpeg, png
                    format", accept_multiple_files=True,)
details = st.button("Check Details")

for uploaded_file in uploaded_files:
    if details:
        if uploaded_file is not None:

            bytes_data = uploaded_file.read()
            image = Image.open(io.BytesIO(bytes_data))
            st.write("file_name:", uploaded_file.name)
            st.image(image, width=100)

        else:
            st.write("No Image File is Uploaded")
            break
```

The parameter accept_multiple_files in st.upload() is a Boolean value that allows our application to accept multiple files to be uploaded when set as True. When the parameter is not defined, the function will accept only a single file, which is the default function. See Figure 6-20.

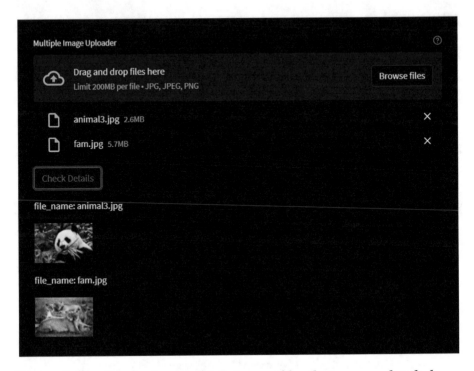

Figure 6-20. *Displaying multiple image files that were uploaded*

To upload multiple files, the same parameter can be used for various file formats that we saw in our previous examples.

Saving Uploaded Documents

In some cases, we need to use the uploaded images for training the model or provide the results associated with them. For this, we need the images to be saved in a directory that can be later reused by our application. We will first get an image using the file_uploader() function and then provide a directory/folder path for it to be saved in. See Listing 6-16.

Listing 6-16. 16_image_save.py

```
# Import following libraries
import streamlit as st
from PIL import Image
import os
import io

# Defining File Upload Method of Streamlit
st.title("Saving File to Directory")
image_file = st.file_uploader("Upload Images",
            type=["png","jpg","jpeg"])

# Defining path where file to be saved
file_save_path = "F:/Books/Apress Streamlit/Chapters/02 Codes/
chapter 6 forms/files"

save_file = st.button("Check Details & Save")
if save_file:

    if image_file is not None:

            # To See details
            image_details = {"file_name":image_file.name,
            "file_type":image_file.type,
                            "file_size":image_file.size}
            st.write(image_details)

            # To View Uploaded Image
            image_data = image_file.read()
            image = Image.open(io.BytesIO(image_data))
            st.image(image, width=250)

            with open(os.path.join(file_save_path,image_file.
            name),"wb") as f:
```

```
            f.write((image_file).getbuffer())

        st.success("Image Saved Successfully")
    else:
        st.write("No Image File is Uploaded")
```

We need to specify file path where the user-uploaded image will be stored. We need to use a relative path for the uploaded image to be saved through our application. The variable file_save_path has been defined in our code. Later the file to be uploaded will be saved when we click the Check Details & Save button in our application, as shown in Figure 6-21.

Figure 6-21. *Saving an image*

We will receive an "Image Saved Successfully" notification once the image is saved, as shown in Figure 6-21, into our specified directory.

In Figure 6-22, the directory shows where the uploaded image has been stored. We can use the same method to store various file documents in the directory.

Figure 6-22. *Displaying the uploaded image file that is now saved in the directory*

Note   If a file with the same name exists in the specified directory or folder, it will override the existing one.

Finally, we will discuss how to submit all the previously mentioned types of form data.

Submit Button

When input form data is changed by the user, the application needs to rerun, causing a bad user experience. To solve this issue of the application rerunning every time the user makes changes in the data, we can use the form_submit_button() function, as shown in Listing 6-17.

Listing 6-17. 17_form_submit.py

```
import streamlit as st

my_form = st.form(key='form')
my_form.text_input(label='Enter any text')

# Defining submit button
submit_button = my_form.form_submit_button(label='Submit')
```

When text is entered in the text box, we will hit the Submit button, as shown in Figure 6-23.

Figure 6-23. *Displaying a Submit button with input form text*

When we hit the Submit button, the application reruns. It is similar to the st.button() function but differs in functionality. The Submit button posts the state of the widgets in batches stored in a form. The st.form_ submit_button should be associated with a form or Streamlit will enter an error state.

Summary

We started this chapter with text boxes in a form. We also saw how a text box can be used to insert a password. We then discussed multiline text inputs using a text area. Later, we discussed taking numbers as input.

We discussed time and date inputs when user wants to submit a form. We also looked at a color selection method where the user can give a color code as input to the form.

Then we discussed different file formats such as .txt, .docx, .csv, and images that can be accepted as input data. We saw how these files can be uploaded and saved in a directory. Finally, we learned how to submit this form data using a submit button.

In the upcoming chapter, we will look at layouts and navigation.

CHAPTER 7

Columns and Navigation

In this chapter, we will discuss how to set up columns, layouts, and navigation into which we can insert elements. These will help to divide our application into grids. This type of navigation is used when we need more than one page in our application. We will also explore how to use sidebars. Let's get started with columns.

Columns

We will be inserting side-by-side columns into our application. We will be using the `st.columns()` function to define columns, which is built in to Streamlit.

The number of columns to be created is specified by an integer value given as a parameter in the `st.columns()` function. See Listing 7-1.

Listing 7-1. 1_columns.py

```python
import streamlit as st

#Defining Columns
col1, col2 = st.columns(2)
```

© Sujay Raghavendra 2023
S. Raghavendra, *Beginner's Guide to Streamlit with Python*,
https://doi.org/10.1007/978-1-4842-8983-9_7

```
# Inserting Elements in column 1
col1.write("First Column")
col1.image("files/fam.jpg")

# Inserting Elements in column 2
col2.write("Second Column")
col2.image("files/fam.jpg")
```

In our example, we have created two columns with an image as element in each one. The columns are equally divided, as revealed in Figure 7-1.

Figure 7-1. *Two columns displaying elements in them with the same size*

We have put two elements in each column; one is text, and the other is an image. The same image has been used to show the difference between the columns, if any. The columns created are responsive in nature, which means when an application is viewed on different devices, the columns change their size accordingly.

Spaced-Out Columns

We can set a column's size by specifying its width. The column with the maximum width is known as a *spaced-out column*. See Listing 7-2.

Listing 7-2. 2_spaced_out.py

```python
import streamlit as st
from PIL import Image

img = Image.open("files/fam.jpg")

st.title("Spaced-Out Columns")

# Defining two Rows
for _ in range(2):
    # Defining no. of columns with size
    cols = st.columns((3, 1, 2, 1))
    cols[0].image(img)
    cols[1].image(img)
    cols[2].image(img)
    cols[3].image(img)
```

In this listing, we have defined four columns with two rows that have images in them (Figure 7-2). The first column is the widest, followed by the third column, and the other two columns have the same size. The first and third columns are said to be *spaced-out* as they are wider compared to the second and fourth columns.

Figure 7-2. *Spaced-out columns*

We can specify any number of columns with spaced-out columns.

Note   A tuple can also be used in the st.streamlit() function to specify the width of each column.

Columns with Padding

Padding is added to create extra space between two columns. This helps to differentiate the elements present in the two columns. Listing 7-3 illustrates how to do this.

Listing 7-3. 3_column_padding.py

```python
import streamlit as st
from PIL import Image

img = Image.open("files/fam.jpg")

st.title("Padding")

# Defining Padding with columns
col1, padding, col2 = st.columns((10,2,10))

with col1:
    col1.image(img)

with col2:
    col2.image(img)
```

We have defined an empty column as padding, as shown in Figure 7-3.

Figure 7-3. *Columns with padding*

Note   Padding is an empty column placed between two columns to create a space there.

Grids

Grids can be created using the st.columns() function as shown earlier. Creating multiple columns in a loop will get us a grid. See Listing 7-4.

Listing 7-4. 4_grid.py

```
import streamlit as st
from PIL import Image

img = Image.open("files/fam.jpg")

st.title("Grid")

# Defining no of Rows
for _ in range(4):
    # Defining no. of columns with size
    cols = st.columns((1, 1, 1, 1))
    cols[0].image(img)
    cols[1].image(img)
    cols[2].image(img)
    cols[3].image(img)
```

Grids are used to align the elements inside them using size and position. In our example, we have defined (4 * 4) rows and columns with the same size, as displayed in Figure 7-4.

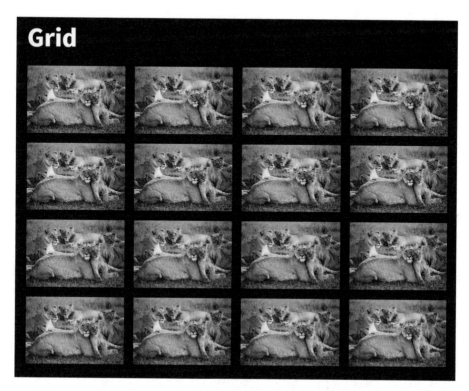

Figure 7-4. *Displaying grids*

Grids can be used for an image gallery or a recommendation system for books, music, movies, etc., where we need to specify a large amount of data in a grid format on a single page.

Expanders/Accordions

When we want to hide additional information from the user or don't want information to always appear on our application, we can use accordions. These are also known as *expanders* as when the user toggles an expander into its open state, it expands and displays the extra information. See Listing 7-5.

Listing 7-5. 5_expanders.py

```python
import streamlit as st

st.title('Exapanders')

# Defining Expanders
with st.expander("Streamlit with Python"):
    st.write("Develop ML Applications in Minutes!!!!")
```

We have created expanders with some text information in them. The text information is shown only when the user clicks the expander. Figure 7-5 shows a closed expander.

Figure 7-5. *Displaying closed expanders*

When you click an expander, you will see the information that is available in it, as in Figure 7-6.

Figure 7-6. *Displaying text information when opened*

The information can be text, image, chart, etc.

Note   We can use the `write()` function to add information in accordions or expanders.

Containers

To insert more than one element, we can use a container in Streamlit. A container is invisible in nature and can be defined with `st.container()`. See Listing 7-6.

Listing 7-6. 6_containers.py

```python
import streamlit as st
import numpy as np

st.title("Container")

with st.container():
    st.write("Element Inside Contianer")

    # Defining Chart Element
    st.line_chart(np.random.randn(40, 4))

st.write("Element Outside Container")
```

We have used a `with` statement in the `st.container()` to insert multiple elements in the container. See Figure 7-7.

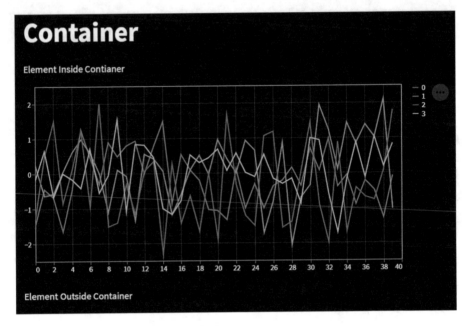

Figure 7-7. *Container with chart objects*

We can insert elements in the container out of order, which can be done by using the `container.write()` function. See Listing 7-7.

Listing 7-7. 7_container_out_of_order.py

```python
import streamlit as st
import numpy as np

st.title("Out of Order Container")

# Defining Contianers
container_one = st.container()
container_one.write("Element One Inside Contianer")
st.write("Element Outside Contianer")

# Now insert few more elements in the container_one
container_one.write("Element Two Inside Contianer")
container_one.line_chart(np.random.randn(40, 4))
```

We have inserted three elements in the container. Two elements are text functions, and one is a line chart, as shown in Figure 7-8.

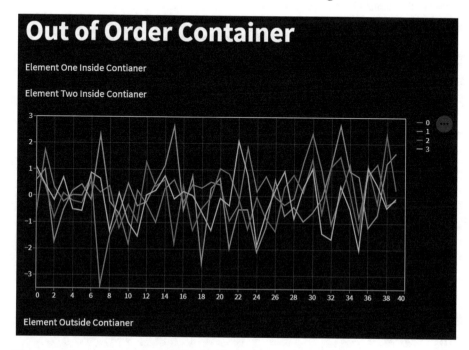

Figure 7-8. Out-of-order container

The order of inserting elements varies and hence is called *out of order*. This helps in creating more flexible applications.

Empty Containers

As the name suggests, the container is empty, and we can insert only one element into it. To define an empty container, we can use the st.empty() syntax. See Listing 7-8.

Listing 7-8. 8_empty_container.py

```python
import streamlit as st
import time

# Empty Container
with st.empty():
    for seconds in range(5):
        st.write(f"⏳ {seconds} seconds have passed")
        time.sleep(1)
    st.write("✔ Times up!")
```

This creates the empty container shown in Figure 7-9.

Figure 7-9. *Displays an empty screen a) before and b) after the time is past*

Sidebars

A sidebar is a pane that is displayed on the side of the application. It allows the user to stay focused on the main content. We will use the `st.sidebar()` function to define a sidebar in our application. See Listing 7-9.

Listing 7-9. 9_sidebar.py

```python
import streamlit as st

# Sidebar
st.sidebar.title("Sidebar")
```

```
st.sidebar.radio("Are you a New User", ["Yes", "No"])
st.sidebar.slider("Select a Number", 0,10)
```

In Listing 7-9, we will be displaying a slider and radio buttons in a sidebar, as shown in Figure 7-10.

Figure 7-10. *Sidebar*

We can use a sidebar to display a navigation link where one can switch from one page to other.

Multipage Navigation

We can create multiple-page apps by using navigation. In Streamlit, the navigation sidebar shows the pages that are created. The pages can be navigated quickly and easily as the front end never reloads. The multipage application is divided into two parts: the main page and the pages.

Main Page

First, we will create a Python file in a directory that acts as the main page. This is similar to the single-page application we created earlier. We can run this file using the following command:

streamlit run file_name.py

We have named Python file `home.py`. The code for `main_page.py` is given in Listing 7-10. Figure 7-11 shows the result.

Figure 7-11. *Displaying the home page*

Listing 7-10. home.py

```
import streamlit as st
# home page
st.title("# Main page 🕸")
st.write("This is Main Page")
```

Figure 7-11 shows the main page. Now we will see how we can add other pages to this `home.py` page.

Pages

We will add multiple Python pages in the folder named `pages` where we have the `main_page.py` file. We will create two files named `page2.py` and `page3.py` in the pages folder.

The structure of the folder is as follows:

```
home.py # It is the main file to run with "streamlit run"
└── pages/ # Folder
    └── page2.py # Second Page
    └── page3.py # Third Page
```

The code for the pages is shown in Listing 7-11 and Listing 7-12.

Listing 7-11. page2.py

```
import streamlit as st
# Second Page
st.title("# Page 2 💥")
st.write("You have navigated to page one")
```

Listing 7-12. page3.py

```
import streamlit as st
# Third Page
st.title("# Page 3 🎖")
st.write("You have navigated to page one")
```

Streamlit takes all the file names of the files that are available in the pages folder and places them in the sidebar of the application, as shown in Figure 7-12 and Figure 7-13.

Figure 7-12. *Displaying the second page*

Figure 7-13. *Displaying the third page*

Note   The multipage navigation works only in Streamlit versions above 1.10.

Summary

In this chapter, we learned how to place objects into columns and grids and learned more about defining containers with objects. We also discussed how to show extra information using expanders.

We learned how to place objects in sidebars separately from the page navigation. Finally, we discussed how to create multiple pages in a Streamlit application by demonstrating the folder structure.

In the next chapter, we will see the control flow and advanced features of the Streamlit application.

CHAPTER 8

Control Flow and Advanced Features

In this chapter, we will discuss the different alert boxes available. Then we will learn about control flow aspects in Streamlit. We will also discuss how to change the default flow of a Streamlit application with the methods provided. Finally, we will learn advanced features including changing the default configuration of Streamlit and improving performance using caching techniques.

Alert Box

In this section, we will review some of the alert boxes that are available in Streamlit. These boxes are used to give certain information to the user in order for them to interact with the application.

st.info( )

This function helps to provide additional information to the user so they can interact with the application.

© Sujay Raghavendra 2023
S. Raghavendra, *Beginner's Guide to Streamlit with Python*,
https://doi.org/10.1007/978-1-4842-8983-9_8

st.warning( )

This will pop up a warning message to the user.

st.success( )

The st.success() function displays a success message. For example, we can use it when form data is submitted correctly.

st.error( )

This is used to display an error message on the application. It can be used when the file upload does not support certain types of files.

st.exception( )

When we want to handle any exception in our application, we can use the st.exception() function and display a notification once the exception is hit, as shown in Listing 8-1 and Figure 8-1.

Listing 8-1. 1_alert.py

```python
import streamlit as st

st.success("Successful")
st.warning("Warning")
st.info("Info")
st.error("Error")
st.exception("It is an exception")
pass
```

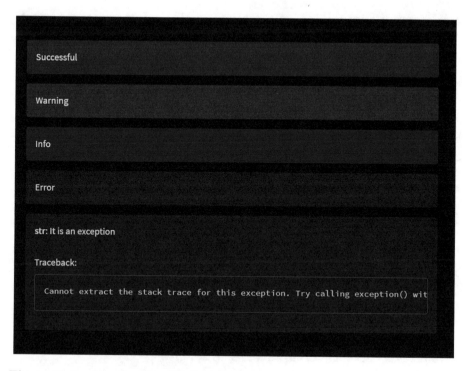

Figure 8-1. *Displaying all alert boxes*

We can use one or more alert boxes in our application, and they can appear in different colors.

Control Flow

We know that Streamlit executes the complete script of an application, but this can be controlled by allowing certain functions to execute, and therefore the default flow of the application can be changed. This can be done by using the following methods.

Stop Execution

We can stop the execution of the script immediately. The command used to stop execution is as follows:

```
st.stop()
```

No other script will run when st.stop() occurs. See Listing 8-2.

Listing 8-2. 2_stop.py

```python
import streamlit as st

name = st.text_input('Text')
if not name:
  st.info('Enter any Text.')
  # Stop function
  st.stop()
st.success('Text Entered Successfully.')
```

In this case, the script will come out of the if condition when any text is entered in the textbox, as we have defined the st.stop() method, and execute the next condition. This changes the default flow of the application.

Rerun the Script

We can rerun the script at any given points of time. To rerun the script, we use the following command:

```
st.experimental_rerun()
```

See Listing 8-3.

Listing 8-3. 3_rerun.py

```
import streamlit as st
import time

st.title('Hello World')
st.info('Script Runs Everytime rerun hits.')
time.sleep(2)

# rerun statement
st.experimental_rerun()
st.success('Scipt Never Runs.')
```

When we hit the rereun() statement in the script, the script runs again from the top, and hence the next script will not be executed.

st.form_submit_button

This button is used to submit all the data entered in the form in batches. When the user clicks the submit button, the script reruns. It is a special type of button that should be associated with forms. See Listing 8-4.

Listing 8-4. 4_submit.py

```
import streamlit as st

sub_form = st.form(key='submit_form')
user_name = sub_form.text_input('Enter your username')

# Submit button associated with form
submit_button = sub_form.form_submit_button('Submit')

st.write('Press submit see username displayed below')

if submit_button:
    st.write(f'Hello!!!! {user_name}')
```

In Listing 8-4, the sub_form is associated with the submit button; hence, when the button is clicked, it will send the data in batches and be displayed on the application. This button can be associated to any of the form widgets that we discussed in Chapter 5.

Advanced Features

In this section, we will discuss the advanced features available in Streamlit. A more detailed description about each feature and its application appears with the code snippets.

Configuring the Page

We can change the default page title and icon used by Streamlit by configuring the page. This page configuration is done by using the st. config_page() method. See Listing 8-5.

Listing 8-5. 5_page_configuration.py

```
import streamlit as st

st.set_page_config(page_title='ML App', page_
icon=':robot_face:')

st.title("Page Configured")
```

The previous code will change the page title and icon of the application, respectively.

st.echo

When we want the code that creates the graphics at the front end, we can use echo. We can also adjust when the code is displayed, i.e., before or after execution of the code. See Listing 8-6.

Listing 8-6. 6_echo.py

```python
import streamlit as st

with st.echo():
    txt = st.text_input('Text')
    if not txt:
        st.warning('Input a text to see sample code.')
        st.stop()
    st.success('Thank you for text input.')
```

In this example, we have displayed the code after execution by getting the text input from the user. Figure 8-2 shows the result.

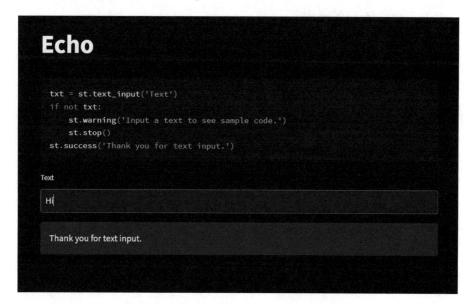

Figure 8-2. *Displaying code using the echo function*

st.experimental_show

To debug our application, we can use the experimental_show() method. We can pass multiple arguments to debug the application. When this method is called, it returns a None value so that the method can be used only for debugging. See Listing 8-7.

Listing 8-7. 7_experimental_show.py

```
import streamlit as st
import pandas as pd
import numpy as np

st.title("Experimental Show for Debugging")
# Dataframe
df = pd.DataFrame(np.random.randint(0,100,size=(5, 4)),
columns=list('WXYZ'))

# Defining Experimental show
st.experimental_show(df)
```

In Listing 8-7, we have used the show method to debug the dataframe created.

Figure 8-3 shows the experimental_show() method displaying the dataframe to be debugged.

Experimental Show for Debugging

df

	W	X	Y	Z
0	26	53	18	32
1	30	83	80	17
2	24	33	59	91
3	33	51	54	38
4	61	80	76	1

Figure 8-3. *Displaying a dataframe given by the experimental_*
show() method

Session State

Session state is a unique feature of the Streamlit library where we can store
values and share the same values between the application reruns. It holds
a key-value pair similar to a Python dictionary. We can use the following
command to access session state:

st.session_state

We can use the key to access the session state by using the following
command:

st.session_stat['key']

Note   The session state holds the value of only the current session,
and after refreshing, a new session state is created.

Once we refresh the page, a new session is created, and the value stored in the session state is erased. The application then stores the value of the new session. See Listing 8-8.

Listing 8-8. 8_session_state.py

```python
import streamlit as st

st.title('Session State')

# Session state initialization
if 'sum' not in st.session_state:
    st.session_state.sum = 0

# Button to add value
add = st.button('Add One')
if add:
    st.session_state.sum += 1

st.write('Total Sum = ', st.session_state.sum)
```

We have a button that adds one value to the total sum when we click it. Every time we click the button, the application reruns and stores the incremented value in the session state. If the session state is not initialized, then the session state value is set to zero.

Figure 8-4. *Displaying the session state sum a) before initialization and b) after initialization*

In Listing 8-8, we saw that the values are stored between the reruns in the session state.

Performance

In this section, we will look into some of the techniques that help to increase performance by increasing the robustness of computations in an application. We will discuss how to increase the performance of the application by using the default settings as well as advanced options.

Caching

When we want to load a large dataset, manipulate data, or perform heavy computational work, we need our application to remain stable and perform as expected. The caching mechanism is provided by Streamlit to handle such things.

When we use a cache in our application as a decorator, Streamlit will look at the following things:

- The input parameters of the called function

- The external variable value used in the function

- The body of the function

- Any function body used inside the cached function

When Streamlit sees these components for the first time in the same order, it will store the values when the script is executed.

Streamlit monitors if there are any changes in the components by using a hashing technique. This is similar to a key-value pair in the memory store wherein the key is a hash and the value is the object passed by reference.

This is a basic example of a cache:

```
import streamlit as st
```

```
import time

# Defining cache
@st.cache

def add(x, y):
    # Function takes 5 secs to run
    time.sleep(5)
    return x + y

x = 10
y = 60
res = add(x, y)

st.write("Result:", res)
```

When the script runs for the first time, we will get five seconds of wait time. After the first run, when the script is executed again, the result is directly returned with no wait time. This is due to the cache we have defined that takes the previous stored data and provides the result. Hence, we can use the cache for large datasets.

st.experimental_memo

In the case of high computations, we use experimental_memo to store data. It holds cache data in key-value format. The command to initialize experimental_memo is as follows:

```
@st.experimental_memo
```

We can use @st.experimental_memo in place of @cache to store the results from heavy computations. It is derived from the same API as the cache.

st.experimental_memo.clear( )

To clear the cache store in `st.experimental_memo`, we will use the following command:

```
st.experimental_memo.clear()
```

The memo can be used to store downloaded data or for dataframe computations or calculations between *n* digits, respectively.

st.experimental_singleton

All users connected to the app share each singleton object. Since they can be accessed by multiple threads at once, singleton objects must be thread-safe. To define a singleton, we will use the following statement:

```
@st.experimental_memo
```

A Streamlit application shares this key-value store with all of its sessions. It works well for storing large singleton objects over sessions such as database connections or sessions for TensorFlow, Torch, or Keras.

st.experimental_singleton.clear

When the singleton is holding the data and needs to be cleared, we use the following function:

```
st.experimental_singleton.clear()
```

Summary

In this chapter, we discussed various alert boxes that can be used to notify the user about certain information. We also discussed how to change the default control flow of the application.

Finally, we discussed in detail the advanced features that are available in the Streamlit application and that are used to enhance performance, reliability, and stability.

In the next chapter, we will develop a natural language processing application from scratch and deploy it.

CHAPTER 9

Natural Language Processing

In this chapter, we will develop a natural language processing (NLP) application from scratch. This application will demonstrate how it works as well as give you the complete instructions for developing an application using Streamlit and thereafter deploying on the Heroku platform.

NLP App Creation

In this section, we will create an NLP application that will predict whether sentiments are positive, negative, or neutral. In this application, we will take the user input and check the sentiment polarity of the sentence.

First, we need to import the necessary Python packages including Streamlit (Listing 9-1).

Listing 9-1. Import Packages

```
import streamlit as st
# NLP Pkgs
from textblob import TextBlob
from nltk.stem.wordnet import WordNetLemmatizer
import re
```

© Sujay Raghavendra 2023
S. Raghavendra, *Beginner's Guide to Streamlit with Python,*
https://doi.org/10.1007/978-1-4842-8983-9_9

To install the package, you can use the `pip` command.

User Input

We have defined the title of our application with a subheader. We have also defined a text area to get user input. The user input is textual data that will be analyzed further. See Listing 9-2.

Listing 9-2. User Input

```
st.title("NLP")

st.subheader("Welcome to our Application")

text = st.text_area("Enter Your Text")
```

Cleaning the Text

Cleaning text is done in different ways. In our case, we will remove any whitespaces, web links, punctuations, and digits from the text entered by the user. We will also apply lemmatization to the text that will give a single word for the inflected forms. See Listing 9-3.

Listing 9-3. Clean User Text

```
#Keeping only Text and digits
text = re.sub(r"[^A-Za-z0-9]", " ", text)
#Removes Whitespaces
text = re.sub(r"\'s", " ", text)
# Removing Links if any
text = re.sub(r"http\S+", " link ", text)
# Removes Punctuations and Numbers
text = re.sub(r"\b\d+(?:\.\d+)?\s+", "", text)

# Splitting Text
```

```
text = text.split()

# Lemmatizer
lemmatizer = WordNetLemmatizer()
lemmatized_words =[lemmatizer.lemmatize(word) for word in text]
    text = " ".join(lemmatized_words)
```

Predictions

The cleaned text is processed when the Analyze button is clicked. Each word in a given sentence is provided with a value in the TextBlob() function to predict the sentiment. The value of each word is taken into account to determine whether the text is positive, negative, or neutral. When the value of the text is more than zero, then the text is a positive sentiment, whereas if the value is less than zero, then the text is predicted as a negative sentiment. If the text value is zero, then the text corresponds to a neutral sentiment. See Listing 9-4.

Listing 9-4. Predictions

```
if st.button("Analyze"):
        blob = TextBlob(text)
        result = blob.sentiment.polarity
        if result > 0.0:
            custom_emoji = ':blush:'
            st.success('Happy : {}'.format(custom_emoji))
        elif result < 0.0:
            custom_emoji = ':disappointed:'
            st.warning('Sad : {}'.format(custom_emoji))
        else:
            custom_emoji = ':confused:'
            st.info('Confused : {}'.format(custom_emoji))
        st.success("Polarity Score is: {}".format(result))
```

Listing 9-5 shows how to create an NLP application with just a few lines of code in Streamlit.

Listing 9-5. nlp.py

```python
import streamlit as st
# NLP Pkgs
from textblob import TextBlob
from nltk.stem.wordnet import WordNetLemmatizer
import re

def main():
    st.title("NLP")

    st.subheader("Welcome to our Application")

    text = st.text_area("Enter Your Text")

    #Text Cleaning
    #Keeping only Text and digits
    text = re.sub(r"[^A-Za-z0-9]", " ", text)
    #Removes Whitespaces
    text = re.sub(r"\'s", " ", text)
    # Removing Links if any
    text = re.sub(r"http\S+", " link ", text)
    # Removes Punctuations and Numbers
    text = re.sub(r"\b\d+(?:\.\d+)?\s+", "", text)

    # Splitting Text
    text = text.split()

    # Lemmatizer
    lemmatizer = WordNetLemmatizer()
    lemmatized_words =[lemmatizer.lemmatize(word) for word
in text]
    text = " ".join(lemmatized_words)
```

```
    if st.button("Analyze"):
        blob = TextBlob(text)
        result = blob.sentiment.polarity
        if result > 0.0:
            custom_emoji = ':blush:'
            st.success('Happy : {}'.format(custom_emoji))
        elif result < 0.0:
            custom_emoji = ':disappointed:'
            st.warning('Sad : {}'.format(custom_emoji))
        else:
            custom_emoji = ':confused:'
            st.info('Confused : {}'.format(custom_emoji))
        st.success("Polarity Score is: {}".format(result))

if __name__ == '__main__':
    main()
```

Figure 9-1 shows what the application looks like.

Figure 9-1. *Displaying the NLP application*

We will first test the application by entering different input text and predicting the sentiments for the text. The results will pop up, as shown in Figure 9-2.

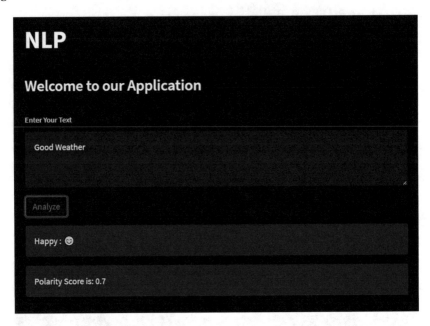

Figure 9-2.   *Displaying a positive sentiment*

In the next section, we will see how to set up files that are later uploaded to a GitHub repository for our application.

Setting Up Files

We will create a few files that are required by Heroku to run our application. The files created are discussed in the following sections.

Requirement Text

We need to create a requirement.txt file (case sensitive). This file will contain all the Python libraries used in our application with its different versions.

There are two ways to create requirement files; one is to list all the libraries used manually. The second way is to install the pipreqs module that automatically creates the requirement.txt file.

We can install pipreqs by using pip as follows:

```
pip install pipreqs
```

After installation, we need to move to our application directory and enter the following command at the command prompt:

```
pipreqs ./
```

Listing 9-6 shows what the requirement.txt file looks like.

Listing 9-6. requirement.txt

```
nltk==3.7
streamlit==1.12.2
textblob==0.17.1
```

Note   Heroku will install all the Python dependencies from requirement.txt.

---

setup.sh

We need to create a shell file named setup.sh (case sensitive) in our application directory where we will add the commands; see Listing 9-7.

Listing 9-7. setup.sh

```
mkdir -p ~/.streamlit/echo "\
[server]\n\
headless = true\n\
port = $PORT\n\
enableCORS = false\n\
\n\
" > ~/.streamlit/config.toml
```

Add the previous command and save it in a shell extension; otherwise, Heroku will give an error message during deployment.

Note   Save the file name as stated in the commands.

Procfile

Finally, we will create a file named Procfile. We need the command shown in Listing 9-8 in our Procfile.

Listing 9-8. Procfile

```
web: sh setup.sh && Streamlit run app.py
```

In this command, web: specifies that it is a web application. The command is specified to run the Heroku application. Finally, app.py is the name of the Streamlit application. Check the name before saving it in a shell file as it may vary in your case.

Now review the files and their names so that there will be no errors during deployment. Check the requirement.txt file if any of the library is added or removed from the application.

GitHub Repository Creation

There are various ways we can create a Git repository. In our case, we will use one of the easiest methods to deploy an application on Heroku.

Note   Heroku requires Git to run as mentioned in the documentation.

We will use the following command to initialize the Git repository from our root project directory in our command prompt:

```
Git init
```

After Git initialization, we will log in from Heroku, as discussed in the next section.

Heroku

Now that we have created all the files required, we will set up Heroku for our application. Log in to the Heroku account from the command prompt by using the following command:

```
Heroku login
```

The command prompt will return the message shown in Figure 9-3.

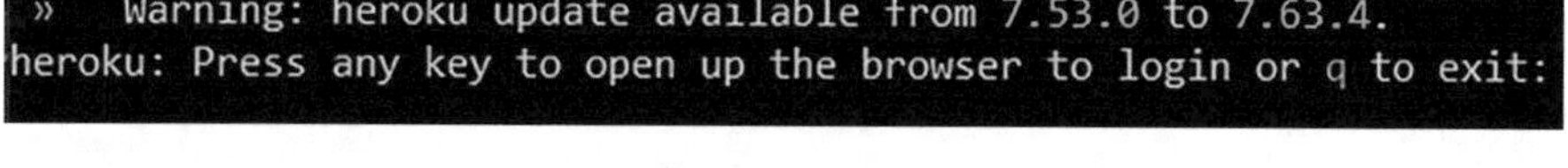

```
»   Warning: heroku update available from 7.53.0 to 7.63.4.
heroku: Press any key to open up the browser to login or q to exit:
```

Figure 9-3. *Displaying the Heroku login message*

Once we press any key, the Heroku login screen will open in our default browser, as shown in Figure 9-4.

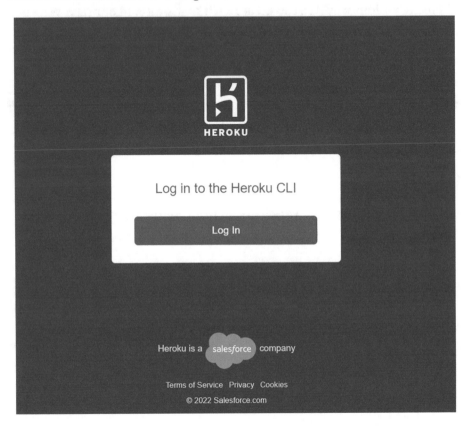

Figure 9-4. *Displaying the Heroku login screen in the browser*

We will click the Log In button and enter our credentials. Once we do that, we will see the Logged In message, as in Figure 9-5.

Figure 9-5. *Displaying the Heroku logged-in message in the browser*

Deployment

In this section, we will learn how to deploy our application. First, we will create an instance for our application by using the following command:

```
heroku create nlp-stream
```

nlp-stream is the name of our application that will be deployed by Heroku.

Note   Duplicate names are not allowed.

We need to use a few commands to push the code to the created instance.

```
git add .
git commit -m "some message"
git push heroku master
```

Once we use the `git push` command, we will notice that it automatically downloads all the Python libraries mentioned in `requirements.txt` and the Python app. After installation, we will see a URL, as shown in Figure 9-6, which means our application has been successfully deployed.

```
remote: -----> Launching...
remote:        Released v3
remote:        https://nlp-stream.herokuapp.com/ deployed to Heroku
remote:
```

Figure 9-6.   Displaying the Heroku deployment

Copy the URL and paste it in a browser to see our app running on Heroku, as shown in Figure 9-7.

Figure 9-7. *Displaying the Heroku deployment*

If there are any changes to the files, we can use the same commands that we used after the Heroku login method. We will get the same URL after the deployment.

Note   The URL we get from Heroku will have the application name to which we have created an instance.

Summary

In this chapter, we developed a Streamlit NLP application with just a few lines of code. The application has the components and features that we learned about in earlier chapters. In this chapter, we also learned how to deploy the Streamlit application to Heroku. Similarly, we can create any application of our choice and deploy it on the cloud. In our final chapter of this book, we will create a Streamlit application with computer vision aspects.

CHAPTER 10

Computer Vision in Streamlit

In this final chapter of the book, we will see one small example of a computer vision application that can predict what animals are in an image. This application will deploy our trained model to classify animal images.

Installing Libraries

In this section, we will install the libraries needed to get started with the ML application.

Install the following libraries using the `pip` command:

```
streamlit
cv2
numpy
tensorflow
```

Now we will import these libraries into our application by employing the code shown in Listing 10-1.

Listing 10-1. Import Libraries

```
import streamlit as st
import cv2
```

© Sujay Raghavendra 2023
S. Raghavendra, *Beginner's Guide to Streamlit with Python*,
https://doi.org/10.1007/978-1-4842-8983-9_10

```
import numpy as np
import tensorflow as tf
from tensorflow.keras.preprocessing import image
from tensorflow.keras.applications.mobilenet_v2 import
MobileNetV2, preprocess_input as mobilenet_v2_preprocess_input
```

Model Deployment

We need to deploy our trained model in our application. This is done by
using the TensorFlow function shown in Listing 10-2.

Listing 10-2. Deploy Model

```
# Deploy Model
with st.spinner('Loading Model...'):
    model = tf.keras.models.load_model("model/mdl_wts.hdf5")
```

The model is in hdf5 format, and it is named animals and stored in the
model folder. Once the model is deployed, then we can upload images. We
will display a spinner to the user while the model is loading. We can deploy
any trained ML model.

Upload Image

We will now take input from the user by uploading an image. This image
can be in JPG or PNG format. See Listing 10-3.

Listing 10-3. Upload Image

```
# Upload file
uploaded_file = st.file_uploader("Choose an image file",
type=["jpg","png"])
```

Map Image Classes

We are mapping five different classes of animals in our dictionary. Later the animal is predicted based on these five classes defined for the model. See Listing 10-4.

Listing 10-4. Map Image

```
# map image classes
animal_dict = {0: 'dog',
          1: 'horse',
          2: 'elephant',
          3: 'butterfly',
          4: 'chicken',
          5: 'cat',
          6: 'cow'}
```

Next, we will apply imaging techniques for the uploaded images.

Apply Imaging Techniques

First, we will convert our image into a byte array object. The same image is then read and converted to RGB format. Later the image is resized to 224 * 244 so that it can be given to the model. See Listing 10-5.

Listing 10-5. Imaging Technique

```
if uploaded_file is not None:

    # Apply Imaging Techniques for uploaded image
    img_bytes = np.asarray(bytearray(uploaded_file.read()),
dtype=np.uint8)
```

```
img = cv2.imdecode(img_bytes, 1)
img = cv2.cvtColor(img, cv2.COLOR_BGR2RGB)
img_resize = cv2.resize(img,(224,224))
```

We have also displayed the uploaded image in the application as follows:

```
# Display Image
st.image(img, channels="RGB")
```

Model Preprocessing

The resized image is given as input to mobilenet. The mobilenet preprocesses the image using its built-in script. See Listing 10-6.

Listing 10-6. Processing Model

```
img_resize = mobilenet_v2_preprocess_input(img_resize)
```

We are also adding an axis to the image predictions as the model had trained with no dimensions added to it.

```
img_reshape = img_resize[np.newaxis,...]
```

Predictions

We add a button named Predict Animal to predict the animal from the model by mapping to the classes defined in the dictionary. See Listing 10-7.

Listing 10-7. Predictions

```
# Button for Prediction
predict = st.button("Predict Animal")
    if predict:
        prediction = model.predict(img_reshape).argmax()
        st.title("Predicted Animal in an image is {}".
format(animal_dict [prediction]))
```

Complete Code

Listing 10-8 gives the complete code for the image classification.

Listing 10-8. cv.py

```
import streamlit as st
import cv2
import numpy as np
import tensorflow as tf
from tensorflow.keras.preprocessing import image
from tensorflow.keras.applications.mobilenet_v2 import
MobileNetV2,preprocess_input as mobilenet_v2_preprocess_input

# Deploy Model
with st.spinner('Loading Model...'):
    model = tf.keras.models.load_model("model/mdl_wts.hdf5")

# Upload file
uploaded_file = st.file_uploader("Choose an image file",
type=["jpg","png"])

# map image classes
animal_dict = {0: 'dog',
            1: 'horse',
```

```
                2: 'elephant',
                3: 'butterfly',
                4: 'chicken',
                5: 'cat',
                6: 'cow'}

    if uploaded_file is not None:

    # Apply Imaging Techniques for uploaded image
    img_bytes = np.asarray(bytearray(uploaded_file.read()),
    dtype=np.uint8)
    img = cv2.imdecode(img_bytes, 1)
    img = cv2.cvtColor(img, cv2.COLOR_BGR2RGB)
    img_resize = cv2.resize(img,(224,224))

    # Display Image
    st.image(img, channels="RGB")

    img_resize = mobilenet_v2_preprocess_input(img_resize)
    img_reshape = img_resize[np.newaxis,...]

    # Button for Prediction
    predict_img = st.button("Predict Animal")
    if predict_img:
        prediction = model.predict(img_reshape).argmax()
        st.title("Predicted Animal in an image is {}".
        format(animal_dict [prediction]))
```

Figure 10-1 shows our application classifying animals by taking input from the user.

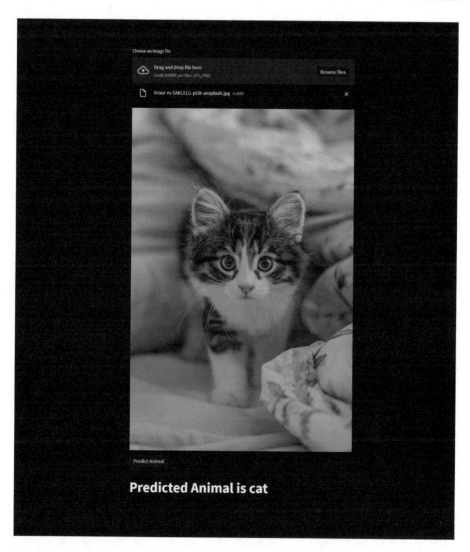

Figure 10-1. *Displaying a computer vision application*

Index

A

Alert box in Streamlit
    st.error() function, 162
    st.exception() function, 162, 163
    st.info() function, 161
    st.success() function, 162
    st.warning() function, 162
Altair
    area graph, 62, 63
    boxplot, 61, 62
    heatmap, 63, 64
    statistical visualizations, 60
Animated snowflake, 96
Area graph/chart, 51, 52, 62, 63
Audio file, 90–92
Audio formats, 92

B

Background image, 87, 88
Balloons, 95
Bar charts, 48, 49, 65, 73
bar() function, 72
Bar graph, 71–75
Boxplot, 61, 62
Built-in visualization functions
    area chart, 51, 52
    bars, 48, 49

    line chart, 50
    map, 53, 54
Buttons
    check box, 102–104
    create, 99
    download, 108, 109
    drop-down, 104–106
    multiselect output, 106, 107
    preselect, checkbox output, 104
    progress bars, 109, 110
    radio buttons, 100–102
    spinner, 111

C

Caching mechanism, 171, 172
Captions, 21–23, 82
Check boxes, 102–104
Code, 26–28
Columns, 50, 76, 145–146
    with padding, 148, 149
    spaced-out column, 146–148
Computer vision application, 195
Computer vision in Streamlit
    complete code, 193, 194
    imaging techniques, 191, 192
    import libraries, 189
    install libraries, 189

© Sujay Raghavendra 2023
S. Raghavendra, *Beginner's Guide to Streamlit with Python*,
https://doi.org/10.1007/978-1-4842-8983-9

Printed in the United States
by Baker & Taylor Publisher Services